Matthias Merten

Interactive Service Robots

AF145467

Matthias Merten

Interactive Service Robots

System Design based on Systems Engineering and Decision Making Methods

Südwestdeutscher Verlag für Hochschulschriften

Impressum / Imprint

Bibliografische Information der Deutschen Nationalbibliothek: Die Deutsche Nationalbibliothek verzeichnet diese Publikation in der Deutschen Nationalbibliografie; detaillierte bibliografische Daten sind im Internet über http://dnb.d-nb.de abrufbar.
Alle in diesem Buch genannten Marken und Produktnamen unterliegen warenzeichen-, marken- oder patentrechtlichem Schutz bzw. sind Warenzeichen oder eingetragene Warenzeichen der jeweiligen Inhaber. Die Wiedergabe von Marken, Produktnamen, Gebrauchsnamen, Handelsnamen, Warenbezeichnungen u.s.w. in diesem Werk berechtigt auch ohne besondere Kennzeichnung nicht zu der Annahme, dass solche Namen im Sinne der Warenzeichen- und Markenschutzgesetzgebung als frei zu betrachten wären und daher von jedermann benutzt werden dürften.

Bibliographic information published by the Deutsche Nationalbibliothek: The Deutsche Nationalbibliothek lists this publication in the Deutsche Nationalbibliografie; detailed bibliographic data are available in the Internet at http://dnb.d-nb.de.
Any brand names and product names mentioned in this book are subject to trademark, brand or patent protection and are trademarks or registered trademarks of their respective holders. The use of brand names, product names, common names, trade names, product descriptions etc. even without a particular marking in this works is in no way to be construed to mean that such names may be regarded as unrestricted in respect of trademark and brand protection legislation and could thus be used by anyone.

Coverbild / Cover image: www.ingimage.com

Verlag / Publisher:
Südwestdeutscher Verlag für Hochschulschriften
ist ein Imprint der / is a trademark of
OmniScriptum GmbH & Co. KG
Heinrich-Böcking-Str. 6-8, 66121 Saarbrücken, Deutschland / Germany
Email: info@svh-verlag.de

Herstellung: siehe letzte Seite /
Printed at: see last page
ISBN: 978-3-8381-3503-8

Zugl. / Approved by: Ilmenau, TU, Diss., 2012

Copyright © 2013 OmniScriptum GmbH & Co. KG
Alle Rechte vorbehalten. / All rights reserved. Saarbrücken 2013

Abstract

Robots successively take over tasks to simplify the life of humans. Already now, robots transport goods in storage buildings; industrial robots manufacture cars; or service robots clean floors in apartments. In the near future, in particular interactive service robots, which communicate with humans, recognize people, or response to natural language, are expected to significantly change our life. The field of application of such service robots ranges from small systems for entertainment, to complex systems guiding visitors in large buildings or interactively assisting people in their homes. Several complex interactive service robots have already been developed and operate as example installations taking over guidance tasks or serving as home assistants. However, none of these systems have become an off-the-shelf product or have achieved the predicted breakthrough so far. The challenges of the design of such systems are, on the one hand, the combination of cutting edge technologies to a complex product; on the other hand, the consideration of requirements important for the later marketing during the design process.

In the framework of this book, two interactive service robot systems are developed that have the potential to overcome current market entry barriers. These robots are designed to operate in two different environments: one robot guides walked-in users in large home improvement stores to requested product locations and interacts with the customer to provide product information; the other robot assists elderly people to stay longer in their homes and takes over home-care tasks. This work describes the realization of the embedded systems of both robots. In particular, the design of low-level system architectures, energy management systems, communication systems, sensor systems, and selected aspects of mechanical implementations are carried out in this work. Multiple embedded system modules are developed for the control of the robots' functionalities; the development processes as well as the composition and evaluation of these modules are presented in this work.

To cope with the complexity and the various factors that are important for the design of the robots, this work applies and further develops system engineering methods. The development process is based on the V-Model system design method.

The V-Model helps to structure the design process under consideration of all system requirements. It involves evaluation procedures at all design levels, and thus increases the quality and reliability of the development outputs. To support design decisions, this work proposes to combine the V-Model with the Analytic Hierarchy Process (AHP) method. The AHP allows to evaluate technical alternatives for design decisions according to overall criteria, a system has to fulfill. This work defines seven criteria that characterize a service robot: Adaptability, Operation Time, Usability, Robustness, Safeness, Features, and Costs. These criteria are weighted for each individual robot application. The AHP allows to evaluate technical design alternatives based on the weighted criteria to reveal the best technical solution. The integration of the AHP into the V-Model development is tested and improved during the design process of the shopping robot system. The generality of this combined systematic design approach is validated during the design of the home-care robot system.

The resulting shopping robot is successfully used in shopping and guidance applications in stores, exhibitions, and trade fairs. The home-care robot has been applied successfully during user trials and will be introduced to the market in 2012. This book demonstrates the combination of the V-Model and the AHP as a highly beneficial method for the development of complex interactive robots. The defined criteria are generally applicable to service robot systems. A modification or definition of further criteria would allow for a transfer of this system design approach to other complex systems.

Acknowledgments

First of all, I want to thank my supervisor Professor Horst-Michael Gross for giving me the opportunity to work in the exciting field of interactive mobile service robots. It was always fascinating to see robots growing up from purely technical systems to useful robotic assistants in his laboratory; and a pleasure to help enlarging his robot family. I am also grateful to all other members of the Neuroinformatics and Cognitive Robotics Lab for contributing to this work by inspiring discussions.

I would like to acknowledge the debt I owe to all my colleagues at the MetraLabs GmbH. It is a team of great engineers creating an inspiring work environment. In particular, I would like to thank Andreas Bley, Christian Martin, and Norbert Herda, who developed service robots with me from the very beginning. To be part of such a team allowed me to focus on the field of embedded systems design of the robots.

My final thanks are addressed to those people who supported me outside the world of robotics. In particular my parents, for always believing in me; and Katharina, for cheering me up, pushing me forward, and listening to all my thoughts. Without her support, the way to finish this work would have been much longer.

To Katharina

Contents

Chapter 1

Introduction

The idea of human-like creatures that accompany and support people fascinated the human race from the beginning of civilization. For example, Leonardo da Vinci (in 1495) or Jacques de Vaucanson (in 1739) already combined their technological knowledge and developed complex automations aiming to create artificial creatures. However, these developments remained on the experimental level until the time of industrialization (second half of the 18th century). The word 'robot' is derived from the Czech word *robota*, which means work or labor and was first mentioned in the play R.U.R. (Rossum's Universal Robots) by the writer Karel Čapek (1920).

Today, the word 'robot' is used for different kinds of machines that are controlled by computer programs. These machines can be classified into three groups: mobile robots, industrial robots, and service robots. Mobile robots move within their environment, e.g., Automatic Guided Vehicles (AGVs) that drive autonomously along markers or wires on the floor. Industrial robots are usually manipulators that support manufacturing processes and are installed on fixed positions. The definition in the field of service robots is ambiguous. The International Federation of Robotics (IFR) [IFR, 2010] has proposed a definition for a service robot as follows:

> A service robot is a robot which operates semi- or fully autonomously to perform services useful to the well-being of humans and equipment, excluding manufacturing operations.

The present work focuses on the field of service robots and will describe the systematic development process of mobile platforms for public and home applications.

In the year 2004 (the beginning of this work), more than 1.3 million service robots were in use [IFR, 2004]. These robot systems were mainly used as floor cleaning robots (ca. 44 %), toys (ca. 44 %), lawn movers (ca. 3 %), and robots for education and training (ca. 1 %). Only a small number of 15 robots acted in public relations as guiding robots, marketing robots, or hotel and food preparation robots. Some examples of such installations are:

- Museum guide *Rhino* in the Deutsche Museum Bonn, 6 days, 1997 [Burgard et al., 1999],

- Three museum guides in the Museum für Kommunikation Berlin, since 2000 [Graf and Barth, 2002],

- Eleven exhibition guides *RoboX* at the Swiss National Exhibition Expo.02, 5 months, 2002 [Siegwart et al., 2003],

- Two marketing robots *Mona* and *Oskar* at the Marken- und Kommunikations- zentrum der Adam Opel AG in Berlin, 12 months, 2003 [Opel, FHG, 2010].

Most of these applications were temporal installations for demonstration and testing. The customer's benefit of these applications - greeting and guiding of visitors through an exhibition - was low compared to the purchase and operating costs of the robots. None of these systems were certified under legal industrial laws and did not become an off-the-shelf product.

The elaboration of more customer needs would possibly allow for a wider usage of such systems. Service robots might carry out customer consultancy, product adver- tisement, store guidance, or customer surveys. However, such functionalities add an enormous complexity to the system requirements. The design of software, hard- ware, mechanical parts, robot-user-interactions, and the robots' appearance have to be considered under increased demands of robustness, acceptance, usability, and costs. Therefore, systematic approaches are required for the development process of such complex systems.

The purpose of this book is to present the development of interactive service robots under application of systematic design methods. This work describes the realization of the embedded systems (e.g., control architectures, power supply systems, sensors systems, and interacting systems) of service robot platforms for shopping and home-care applications. For a systematic design approach, this work proposes and investigates a combination of a V-Model design process [V-Model XT, 2009] with the Analytic Hierarchy Process (AHP) method, which supports the decision processes [Saaty, 1994]. This work defines criteria that characterize a service robot and evaluates various technical alternatives. The AHP is applied to reveal the best technical solutions based on the weighted criteria of a specific robot system. The systematic design approach was tested and improved during the design process of a shopping robot system. This robot is now successfully used in shopping and guidance applications in stores, exhibitions, and trade fairs. The adaptability of the developed systematic design approach was validated during the design of a home-care robot system. This second robot has been applied successfully during user trials and will be introduced to the market in 2012. Both service robot developments demonstrated the successful application of the proposed system design method to the complex design process of interactive service robots.

1.1   Timeline of the Service Robot Developments

The work described in this book started with the development of a shopping robot platform for home improvement stores. Such stores have big sales areas and market a large variety of products. The responsibility of a robot in such stores would be to interact with customers and assist during shopping. The robot should be able to guide customers to requested products, give pricing and product information, or show and advertise services of the store. The benefit of a shopping robot application is that the customer would find faster the requested products and would be provided with higher service quality, because the employees could concentrate on sale conversations with higher needs of consultancy. Additionally, the store could acquire statistical information about requested products to further adapt their services and to increase the turnover.

The development process was carried out in several research projects and on multiple robot platform versions (Figure 1.1).

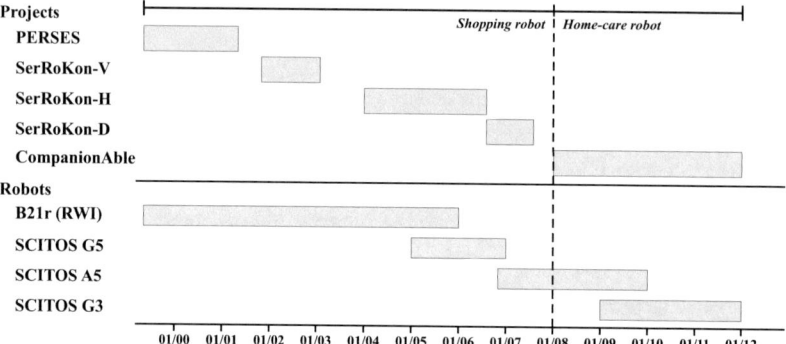

Figure 1.1: *Overview of relevant projects and applied robot platforms for the development of the shopping robot application until the end of 2007 and the home-care robot from beginning of 2008.*

The development of the interactive shopping robot started with the *PERsonenlokalisation und PERsonentracking für mobile SErvicesysteme (PERSES)* project. This project was carried out by the Neuroinformatics and Cognitive Robotics Lab at the Ilmenau University of Technology from April 1999 to March 2001. It was mainly research oriented and aimed to develop a shopping robot that guides users to dedicated products, follows customers to provide additional information, and to communicate with users based on visual and acoustic interaction [Gross and Böhme, 2000]. All tests and developments within the *PERSES* project were based on a research robot platform B21r from the company RWI. This platform was sufficient for the usage in research activities but too complex and expensive for real-life applications. Other systems available at this time were inflexible and technologically not applicable for the shopping application.

There was a clear need for a new robot platform suitable for professional usage. This

4

SCITOS G5
(Prototyp I)

SCITOS A5

SCITOS G3

Figure 1.2: *Robot systems used in the SerRoKon and CompanionAble projects: Prototype of the research platform SCITOS G5 developed within the project, shopping robot SCITOS A5 used for field tests, and SCITOS G3 developed for home environments.*

need was addressed by the three *Service-Roboter-Konzeption (SerRoKon)*[1] projects with the goal to develop a shopping robot based on a self-designed robot platform. The *SerRoKon-V* project, which was carried out from October 2001 until January 2003, specified the final application, defined required algorithms, and compiled a specification of an adequate robot platform.

This work started with the *SerRoKon-H* project, January 2004 to June 2006, and focused on the realization of the new robot system. The first part of this project included further improvements of software algorithms from the *PERSES* project, the research for an adequate platform architecture, and the design of the robot prototype *SCITOS G5* (Figure 1.2). In the second phase, the improved software modules were adapted to the first prototype of the robot. The third part of the

---

[1]The projects were financed by the Thuringian Ministry of Science, Research and Culture, FKZ B509-03007 (SerRoKon-H, 2004-2006) and the Thuringian Ministry of Economy, Technology and Labor (TMWTA), Development Bank for Thuringia (TAB), FKZ 2006-FE-0154 (SerRoKon-D, 2006-2007)

SerRoKon-H project was concerned with the analysis of design mistakes and the compilation of a revision list for a redesign of the first robot prototype.

The project *SerRoKon-D*, July 2006 to June 2007, was installed to demonstrate and to evaluate the results of the previous developments under real-life conditions. Three of the new robot systems, named *SCITOS A5*, were installed to operate completely autonomously in a home improvement store in Bavaria, Germany by March 2008 [Gross et al., 2008]. A second evaluation period was added after the ending of the *SerRoKon-D* project to further optimize and test the systems. During this one year period, ten robot systems were installed in three stores. The robots drove 2,187 km in ten months and successfully guided around 8,600 customers to chosen products [Gross et al., 2009].

The *SerRoKon* projects were mainly carried out by one research partner and two industrial partners. The Neuroinformatics and Cognitive Robotics Lab at the Ilmenau University of Technology as the research partner in the project was responsible for the development of the robots' intelligence. This includes navigation, localization and path planning, the architecture of the robots' software system, and interaction algorithms for the communication with users. The company MetraLabs GmbH joined the projects in 2003 and was responsible for the development of the physical robot, including the mechanical design, the electronic and sensor modules, the embedded system architecture, and low-level software functionalities. MetraLabs was interested in the manufacturing of the robots in a later phase and had, therefore, the responsibility to develop the system under low cost and production aspects. This work arose from contributions to both partners - the Neuroinformatics and Cognitive Robotics Lab and the company MetraLabs - because of its connectivity to higher software levels as well as to the physical robot (see Section 1.2). The company Toom BauMarkt GmbH, as the second industrial partner, was interested in the usage of the shopping robot. Toom BauMarkt defined the system requirements for a successful integration of the robots in the stores and provided the test markets for the later field experiments.

Following the *SerRoKon* projects, the Neuroinformatics and Cognitive Robotics Lab at the Ilmenau University of Technology and the company MetraLabs participated in the *CompanionAble* project[2] carried out by 18 European partners since January 2008 [Companionable, 2011]. The focus of this project was the development of assistive technologies for care-recipients and care-givers based on a robot companion in combination with a smart home. The robot was designed to drive in the care-recipients home and to offer services that require mobility (e.g., detection of a fall of the care-recipient or user monitoring). Similarly to the previous projects, the Neuroinformatics and Cognitive Robotics Lab was mainly responsible for the development of the robots' intelligence, whereas MetraLabs focused on the design of the robot platform. The outcome of this work was the home-care robot platform *SCITOS G3* (Figure 1.2), which offers a different set of functionalities compared to the shopping robot. This enabled the present work to apply and verify the usability of the proposed system design method to the development process of a robot system for home environments. This demonstrates the generalization of the system design method to another complex system.

1.2   Contribution to the Robot Developments

The contribution of this work to both service robots was the systematic development of the robots' embedded systems. This work addressed the design of the low-level system architectures, the energy management system, the communication systems, the sensor systems, and selected aspects of mechanical implementations (overview Figure 1.3). In addition to conceptual work, this contained in particular the development of 16 control modules for the shopping robot, the design of three complex control modules for the home-care robot, and the selection and evaluation of external modules from suppliers.

For the deliberate consideration of all requirements and the individual design criteria of each robot system, this work applied and designed systems engineering and decision making methods. The V-Model as a system engineering approach for complex development processes was applied and adapted. The AHP was used to determine

---

[2]This project was financed by the European Communitys 7th Framework Program (FP7/2007-2013) under grant agreement no. 216487.

optimal design decisions under consideration of weighted criteria for each individual robot system. The combination of both approaches significantly facilitated and improved the development process, which was necessary to successfully realize both robot platforms with the given resources.

The following components of the service robots were contributed by the work described in this book (Figure 1.3):

System architecture: The system architecture affects the robustness, flexibility, and costs of an embedded system as well as the compatibility to off-the-shelf control modules. It depends on the requirements and the later usage of the system.

Energy management: To assure a high availability of a service robot, it is essential to maximize the ratio of working time to charging time. The concepts of the power supply systems were optimized by power consumption of all electronic modules on the one hand, and the available charging energy on the other hand. Further, it includes considerations of the optimal energy storage.

Communication systems: An interactive service robot consists of two types of communication interfaces: the internal communication between system com-

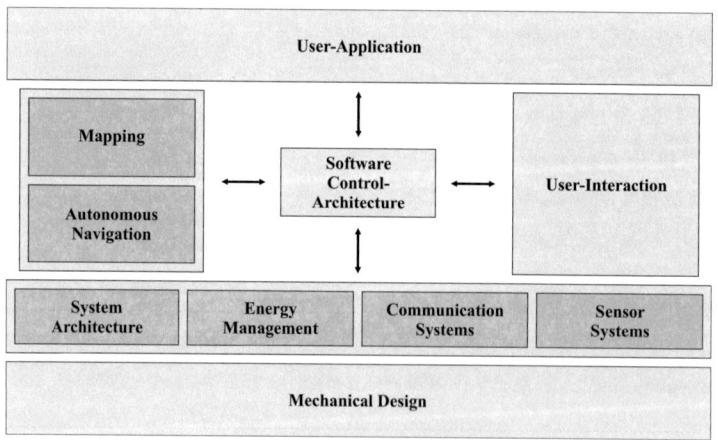

Figure 1.3: *Hardware, software, and mechanical components of a service robot.*

ponents and an external communication with users and enabling systems. The implementation of internal communication systems depends on the system architecture. The main criteria are flexibility, dependability, and costs. The communication to users (Human Machine Interfaces (HMIs)) includes components for information output (e.g., displays, speakers) or input (e.g., touch-sensors, microphones, keys). The interfaces to enabling systems include communication channels to data bases or the Internet.

Sensor systems: Sensors are required to reliably move a robot trough its environment and to detect relevant events in its surrounding area. Therefore, the configuration of the sensor systems depends on the given system environment (e.g., stores, apartments, exhibitions) as well as the final application (e.g., guidance, monitoring). The challenge of sensor systems design is to find an adequate sensor configuration under the consideration of costs.

Mechanical design: The mechanical components of a robot include the drive system, the mechanical framework, the integration of system components, and the overall design. Important aspects considered by this work were the drive system, influencing the movability of the platform and the design, influencing the acceptance of a robot's appearance.

Another main component of a service robot is the software system. This includes algorithms for navigation and localization, user interaction, the user applications, and the higher software control architecture. These components were not part of this work. Nevertheless, this work will discuss some aspects of the software development process that were relevant for the embedded system design processes.

1.3   Content and Structure of this Document

The next chapter of this book gives an overview of the state-of-the-art in interactive service robotics. It describes technical parameters, realizations, and results of real-life applications of comparable robot systems in the fields of shopping and guidance robots as well as home assistant robots. It further presents an interactive service

9

robot developed under systems engineering aspects. At the end of this chapter, advantages and disadvantages of current robot systems are highlighted.

Chapter 3 gives an introduction into system design models. It presents the Waterfall-Model, the Spiral-Model, and the Prototyping and Iterative-Incremental Model. The V-Model, as the basis for the developments of this work, is described in more detail. The adaptation of the V-Model to the development process of interactive service robots is explained.

Chapter 4 introduces the Analytic Hierarchy Process (AHP) and describes the application of this decision method within the V-Model. Main criteria for the characterization of interactive service robot systems are defined. Technical design alternatives (for subsystems like the system architecture or the battery system) that need to be developed as part of every service robot are determined, discussed, and rated using the AHP.

Chapter 5 presents the shopping robot development process based on the combined V-Model and AHP development approach. The system specification is described, i.e., functional requirements, non-functional requirements, and evaluation processes. The system decomposition process is carried out and subsystems, segments, and units are introduced. System characteristics are used to weight the criteria of service robots and conclusions are drawn for the development process using the AHP. The consequences of the AHP decision process are analyzed. The design processes of selected system units are exemplarily described and the system integration process explained. Finally, the integration of the robot system in its dedicated application area including test results is presented.

The following Chapter 6 describes the home-care robot development process, verifying the applicability of the novel design method to further design processes. Similar to the design process of the shopping robot, requirements are derived from the system specification, the decomposition process is illustrated, and the AHP is carried out. Design decisions for the home-care robot are presented compared to the shopping robot. Consequences for the final system properties are discussed.

Finally, Chapter 7 discusses and concludes the results presented in this book.

Chapter 2

State-of-the-art in Service Robotics

This chapter presents the state-of-the-art in service robotics. Because of the vague definition of service robots, this summary focuses on available robot systems with a similar application range than the two systems developed within this work: guidance robots and home-care robots. The following constraints were applied to select the described systems:

- The robot operates in indoor environments.

- The robot is mobile and moves based on wheels.

- The robot's size allows the usage in home or public environments.

- The robot's shape supports the impression of a smart individual.

- The robot serves humans during daily activities.

- The robot interacts with humans in a bidirectional manner.

- The robot's development is completed or it is in the final step of development.

- The robot is intended to be off-the-shelf.

The presented robot systems should give an overview of current applications without the claim of completeness. The description focuses on technical realizations (e.g., drive system, sensor configuration, computing power, battery system). Unfortunately, details of real-life applications are not available for most systems and can just be assumed (see Section 2.4). A robot system for shopping applications that was developed under systems engineering aspects is also presented. At the end of this chapter, the advantages and disadvantages of the systems will be summarizes. Other robot platforms like the *PeopleBot* by MobileRobots [Kuo et al., 2008] or *ARTOS* by the University of Kaiserslautern [Berns and Mehdi, 2010] that could be used for professional applications are not covered by this chapter for reason of clarity.

2.1 Interactive Shopping and Guidance Robots

2.1.1 Museum Robots by Fraunhofer IPA

The Fraunhofer Institute for Manufacturing Engineering and Automation IPA has installed three robot systems (Figure 2.1) in the Museum für Kommunikation Berlin in March 2000 to attract visitors and to demonstrate the possibilities of interactive service robots. The three robots differ by their appearance, behavior, and speech to give every system its own personality [Graf et al., 2000]. *The Inciting* welcomes visitors at the entrance of the exhibition and acts as an entertainer. *The Instructive* gives guided tours in the museum. This robot is equipped with a screen showing videos and pictures. *The Twiddling* is the child of the "robot family", is unable to speak and plays with a ball.

All three robots are based on a similar robot platform. This platform is equipped with a differential drive system with two driven wheels and four castor wheels. The maximum speed of the platform is limited to $1.2\,\mathrm{m/s}$. The sensory configuration of the platform is given by a 2D laser range finder to recognize obstacles and visitors, a gyroscope to compute the movement of the robot, a bumper to signal collisions, several infrared sensors facing upwards to detect persons in the operating area of

Figure 2.1: *Robots in the Museum für Kommunikation Berlin [Fraunhofer IPA, 2011].*

the robot, and emergency stop buttons to disable the drive system. Furthermore, a magnet sensor facing to the ground detects the magnet barriers integrated in the floor and is used to maintain the robots within the defined operating area. The power supply system is able to operate the platform about ten hours.

The operation system of the robots is well adapted for technically unexperienced users Therefore, the user interface consists only of a joystick with two buttons. By this joystick the robot can be set in operation, shut down, or started in different operation modes such as initial localization or self-test. The robot guides its users through the operation modes by speech output [Graf and Barth, 2002].

Since the installation of the robot systems in March 2000, the robots worked daily. Until April 2004, the robots covered a distance of more than 12,700 km without any documented collision with visitors or inventory [Graf et al., 2004]. The robots never left their operating area. This high reliability of the systems was achieved by three factors: The usage of well-established robot technology of the institute, a test period of two months in the museum prior to setting the robots into operation, and a well-defined operating area (compared to the environments of homes or stores).

2.1.2 Mona and Oskar by Fraunhofer IPA

The exhibition robots *Mona* and *Oskar* (Figure 2.2) were also designed by the Fraunhofer Institute for Manufacturing Engineering and Automation IPA in cooperation
with the company GPS GmbH, Germany for the application in the Marken- und
Kommunikationszentrum der Adam Opel AG Berlin [Neuner et al., 2003]. The tasks
of the robots were to inform and to entertain guests in the communication center.
The robots were able to distinguish between single persons and groups to adapt
their speech outputs. *Mona* - the female robot - welcomed visitors at the entrance
and brought them to the exhibition area. *Oskar* - the male robot - took over the
visitors and explained the exhibits. The two robots operated from October 2003
until October 2004 and were tested for two weeks before they were set in operation.
The robots were able to autonomously shut down and restart themselves to reduce
the supervision by the employees. Moreover, functionalities like remote access and
self-diagnostics simplified the usage of the robots for the operator [Graf et al., 2004].

Similar to the robots installed in the Museum für Kommunikation Berlin, *Mona*
and *Oskar* were based on a differential platform with two driven and three castor
wheels. The maximum speed was limited to 0.4 m/s. Due to eight 12 V batteries,

Figure 2.2: *Exhibition robots Mona and Oskar [Fraunhofer IPA, 2010].*

the robots operated up to ten hours. The sensor configuration included two laser range finders, eight ultrasonic sensors placed above the laser range finders, a foam plastic bumper, and two emergency stop buttons. Additional magnet sensors, facing to the floor, detected the edges of the operation area. Both robots were equipped with a touchscreen and loudspeakers to provide visual and audio information to the users.

2.1.3   RoboX by BlueBotics

The swiss company BlueBotics in cooperation with the Autonomous Systems Lab of the Swiss Federal Institute of Technology Lausanne has developed the robot system *RoboX* for the Expo.02 in Switzerland (Figure 2.3). Eleven robots operated about 159 days from May 15 to October 20 at the exhibition [Tomatis et al., 2003]. The goal of this application was to present the current state of robotics technology to the exhibition visitors. The robots aimed to attract the attention of visitors, interacted with them via input buttons and speech output, and gave a pre-defined guided tour.

An exhibition robot has to fulfill three main requirements: First, it must operate autonomously (including self-tests, shut-down, and restart) on a long live cycle. This should minimize the supervision by the exhibition employees. Second, the robot should be self-contained and easily adaptable to other applications. This helps to minimize the effort for installing the systems into the operating place. Finally, the robot should be able to robustly navigate in crowded areas.

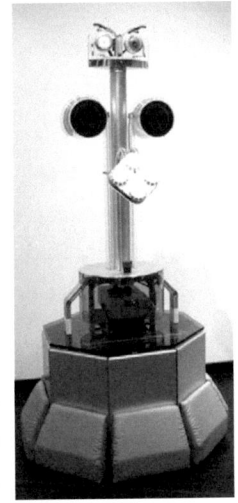

The *RoboX* systems are based on a differential driven platform with one castor wheel at the front and one castor wheel at the backside of the robot. For a robust movement of the four wheeled robot, the back-side castor wheel was mounted on a spring suspension. The robot base has a diameter of 65 cm [Jensen et al., 2002] and consists of

Figure 2.3: *Exhibition system RoboX [Tomatis et al., 2003].*

15

batteries, two control computers, two laser range finders, eight independent bumpers, and an emergency stop button. The upper part of the robot incorporates the interaction system with two independent eyes, mounted on pan-tilt units, two coupled eyebrows, and a simple input device. The left eye is equipped with a small color camera for face detection and tracking, whereas the right eye includes a Light-Emitting Diode (LED) matrix showing symbols and icons. The eyebrows can be tilted to indicate facial expressions. The input device of the robot, in form of four buttons, allows the user to reply to questions by pressing one of these buttons. Another optional input device is an omni-directional microphone array, which can be used for speech recognition. The whole system is about 1.65 m high and is capable of operating up to twelve hours.

The eleven robots at the Expo. 02 drove 3,316 km in 159 days on an operation surface of 320 m^2. The robots served more than 680,000 visitors in a total of 13,313 hours.

2.1.4   Gilberto by BlueBotics

Gilberto (Figure 2.4b) is another example of an interactive service robot by Blue-Botics. It was originally designed to be used at train stations in Italy, but is also intended to be used at airports, museums, exhibitions, trade fairs, and marketing events [Bluebotics SA, 2011a]. The robot has the same platform as *RoboX* (section 2.1.3) and, therefore, similar technical parameters. The main differences are the integrated touch display for user interaction and the adapted design with a height of 2 m and a diameter of 0.8 m. The robot is able to drive with a speed of up to 0.6 m/s and to operate up to eight hours [Bluebotics SA, 2011b].

2.1.5   REEM-H2 by PAL Robotics

The service robot *REEM-H2* (Figure 2.4a), developed by the Spanish company PAL Robotics, was introduced by the end of 2010 [PAL ROBOTICS S.L, 2010a]. This robot was developed to guide and entertain visitors in malls, hotels, exhibitions, airports, or hospitals. The design of the robot incorporates a deposition rack and

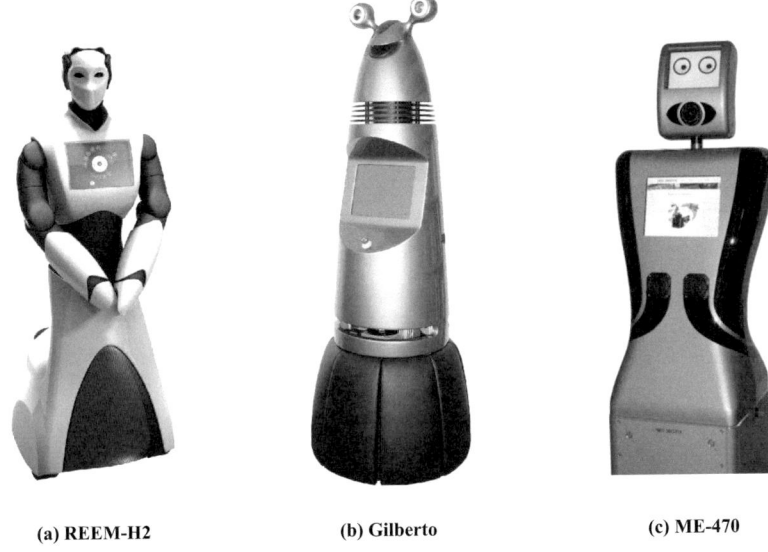

(a) REEM-H2   **(b) Gilberto**   **(c) ME-470**

Figure 2.4: *Guiding robots: (a) Humanoid service robot REEM-H2 by PAL Robotics [PAL ROBOTICS S.L, 2010b], (b) Mobile entertainment robot Gilberto by BlueBotics [Bluebotics SA, 2011a], and (c) Interactive shopping guide ME-470 by Neobotix [GPS GmbH, 2011].*

enables the transport of small objects. The integrated arms can be used to show directions, to gesticulate, or to lift objects up to 3 kg The robot is equipped with software functionalities for autonomous navigation as well as voice and face recognition usable for human like robot interaction. The integrated 12 inch touch display allows for the presentation of multimedia information and users input.

The height of the robot is 1.70 m and its weight 90 kg. It can drive with speeds up to 1.2 m/s. The integrated lithium battery can power the robot up to eight hours. The sensor configuration consists of a laser range finder, ultrasonic sensors, a stereo camera, microphones, accelerometers, and gyroscopes. An embedded PC with an Intel Core-2 Duo processor and an Atom CPU executes software algorithms and user applications [PAL ROBOTICS S.L, 2010b].

2.1.6 ME-470 by Neobotix

The German company Neobotix developed the interactive shopping guide *ME-470* (Figure 2.4c) to attract visitors at exhibitions and trade fairs [GPS GmbH, 2011]. It is designed to show presentations, picture, videos, games, or websites on a 12 inch touch display, located at the upper part of the robot's body. The robot head contains an 8 inch touch display to simulate a robot face and to express emotions. The rotatable head is further equipped with a sound system including microphones and speakers. The casing of the robot can be adapted to the requests of customers. The integration of light effects further increases the attraction of attention.

The robot platform contains a differential drive with two spring-mounted driven wheels at the center line of the robot and four castor wheels. The robot is able to drive up to 1.0 m/s. A 24 V battery with 152 Ah can power the robot for about ten hours. The sensor configuration is given by a laser range finder, four ultrasonic sensors, and a wide angle camera. To execute user software, the robot is equipped with an embedded PC with an 1.8 GHz Atom Dual-Core processor. The height of the robot is 174 cm, the overall weight is about 185 kg.

2.2 Home Assistant Robots

2.2.1 Kompai by Robosoft

The robot *Kompai* (Figure 2.5) was developed by the company Robosoft as a home care system. The robot is intended to support people in scheduling their tasks, reminding to take their medicines, or the creation of shopping lists [Robosoft SA, 2011]. It provides Internet access for email, social networks, or video conferences. Another application is the reception and guidance of visitors at exhibitions and museums. The robot can be controlled remotely or drive autonomously to given positions. Voice

Figure 2.5: *Home care robot Kompai [Robosoft SA, 2011]*

recognition and speech synthesis simplify the user interaction.

The robot platform consists of a differential drive with two driven wheels at the centerline and one castor wheel at the front and one at the back side. The robot base has a footprint of 45 cm by 40 cm and a weight of 25 kg. The sensor configuration of the robot includes a laser range finder, infrared sensors for obstacle and stairway detection, ultrasonic sensors, microphones, and camera systems. For user interaction a tablet PC is attached to the platform.

2.2.2 Care-O-Bot 3 by Fraunhofer IPA

The third generation of the *Care-O-Bot* systems (Figure 2.6) was designed to support the nursing staff at time consuming tasks. The functionalities mainly considered during the development process included transportation and catering. Therefore, the first mission scenario of the robot was to prepare and to serve drinks. This requires the accomplishment of the following tasks: offer a drink on a touch screen, drive to a bar, detect the correct bottle, take the recognized bottle and place it on a tray, open a cupboard, take a glass out of the cupboard and place it on the tray, drive back to the user, and deliver the drink to the user [Reiser et al., 2009b].

Figure 2.6: *Research platform Care-O-bot 3: Left side illustrates first design concepts [Reiser et al., 2009a]. Right side shows the final realization [Care-o-Bot, 2011].*

To be able to fulfill these complex tasks, the *Care-O-Bot 3* was equipped with highly advanced technology. A good maneuverability was enabled by the integration of an omni-directional drive system based on four wheels [Reiser et al., 2009a]. For every wheel, the orientation and driving speed can be set independently; this allows for forward, backward, sideward, and rotation movements. The robot base, with a rectangular shape of 60 cm by 60 cm, consists of a lithium ion battery with 50 V and 60 Ah, a laser range finder, and an embedded PC for navigation purpose. The torso of the robot is equipped with electronic modules and a second embedded PC for robot control. It further includes stereo cameras and a Time of Flight (TOF) camera, which are mounted on a sensor head unit with 5-Degrees of Freedom (DOF) to enable the positioning of the sensors in the direction of interest. The movable cover of the torso can be used to show gestures (e.g., bow, nod, or head shaking). For the manipulation of objects, a light-weight robot arm with 7-DOF is integrated. Additionally, a 7-DOF hand is installed to grab objects. Integrated touch sensors in the fingers allow for a force-controlled grasping of breakable objects, like glasses. Finally, the tiltable tray with an integrated touch screen can be used to transport small objects and to exchange information with users.

2.2.3 Luna by RoboDynamics

Another example of a service robot for the home environment is *Luna* (Figure 2.7), a fully programmable personal robot developed by RoboDynamics. The company aspires to widely distribute the robot to home applications based on two main aspects: an adequate price and an open software architecture [Ackerman and Guizzo, 2011]. The sales price of the robot is planned to be about 1,000 dollar (first versions will be sold for 3,000 dollar). This concept promises extensive sales to private persons. The software architecture is designed to enable an easy creation of new applications that might be available in software stores.

The platform has a diameter of 56 cm, a height of 157 cm, and a weight of 30 kg. The integrated batteries with 12 V and 26 Ah can supply the robot between four and eight hours. The human-robot interaction is based on an 8 inch touch display

RoboDynamics **Luna**   SchultzeWORKS designstudio

Figure 2.7: *Personal robot Luna by RoboDynamics [SchultzeWORKS Design-studio, 2011].*

located in the robot's face. The sensor configuration is given by a high-resolution camera with a digital zoom, a microphone array, and a 3D-sensor. The robot is equipped with a PC including an Atom Dual Core processor.

2.3   Service Robots under Systems Engineering Aspects

The shopping, guidance, and home-care robots described in previous sections are complex systems. Yet, hardly any design process of these systems has been documented and is publicly accessible. One system whose development has been described is the shopping cart robot of the *CommRob* project. Here, this robot will be listed and described, although this system has not been meant to be a real-life application but rather a platform for the research on human-robot interaction in public areas. The design of this autonomous shopping cart aimed to interact with the user in multiple ways (e.g., guidance or product advertisement). The robot platform has been applied for the development of strategies to safely move in dynamic environments, to detect and avoid obstacles, and for robust localization.

Figure 2.8: Shopping cart of the CommRob project [CommRob, 2010].

The development process of this robot system was oriented on the structure of the Waterfall-Model (Section 3.2) in combination with iterative elements. It started with the specification of an initial version of user and system requirements based on the project proposal. These requirements evolved during the course of the project, because of its research character. Considering the specifications (and the experiences from previous developments of the project partners), the system architecture was defined. The decomposition process divided the system into sub-systems, which were realized in the following. Finally, the developed sub-systems were assembled to the complete robot system. The course of this development process was carried out in major iterations (at system level) and minor iterations (at sub-system level), which were arranged to adapt the system requirements and to redesign system components [CommRob, 2009].

The usage of iterations in the linear course of the Waterfall-Model allowed for the consideration of the evolving requirements of this research project. The system decomposition from system to sub-system level made it possible to decrease the system complexity for a more effective design process. However, the development of professional service robots for real-life applications requires major improvements of such a development process. The main issue is that all requirements of the robot system must be identified at the beginning of the project. This is mandatory for a goal-oriented development process under consideration of all relevant functional and technological aspects. It is important to consult customers at an early project for a user-centered development process. An overall quality management is also necessary for a successful development process. The Waterfall-Model provides only limited possibilities to address these aspects. Finally, the decomposition process of complex service robots should include additional steps to decrease the system design complexity and to evaluate the developments at low abstraction levels. Consequently, a design method suitable for highly complex development projects that is applied in professional system developments should be used for the design of interactive service robots. Such a design method is the V-Model that is applied to the service robot developments in this work (Chapter 3.5).

The next section presents an overview of the described robot systems. Because of the research character of the *CommRob* platform, this robot is not considered by this summary.

2.4 Summary

The described robot systems were developed considering different design priorities to achieve a successful integration into guidance applications or home applications (e.g., low system costs of *Luna* or a high user benefit of the *Care-O-Bot 3*). Therefore, the resulting technical parameters of the platforms (summarized in Table 2.1) vary, although the user and system requirements were similar.

Up to now, all of the presented systems are single installations, development platforms, or announced products. None of these systems seems to be in a real-life usage with the perspective to become an off-the-shelf product. The main reason might be seen in a technology-centered rather than an application-centered development process. In this case, the consideration of user and market requirements is inadequate, which leads to unqualified system properties like high system costs (*REEM-H2*, *Care-O-Bot 3*), insufficient user interfaces (IPA museum robots, *RoboX*, *Luna*), unsatisfying designs (*ME-470*, *Kompai*), or inappropriate safety concepts (*REEM-H2*, *ME-470*, *Care-O-Bot 3*, *Kompai*).

Table 2.1: *Summary of technical parameters of the presented robots. Estimated values are written in gray, unknown values are market by a '–'.*

	Museum Robots	Mona and Oskar	RoboX	Gilberto	REEM-H2	ME-470	Kompai	Care-O-Bot 3	Luna
Max. Operation Time [h]	10	10	12	8	8	10	–	–	8
Height [cm]	–	–	165	200	170	174	120	155	157
Weight [kg]	–	–	–	–	90	185	25	–	30
Differential Drive	√	√	√	√	√	√	√		√
Omni-Directional Drive								√	
Driven Wheels	2	2	2	2	2	2	2	4	2
Castor Wheels	4	4	2	2	–	4	2	0	–
Max. Trans. Speed [m/s]	1.2	0.4	0.6	0.6	1.2	1.0	1.0	–	–
Laser Range Finders	1	2	2	2	1	1	1	2	0
Ultrasonic Sensors	0	8	0	0	–	4	9	0	0
Bumper Elements	1	1	8	8	–	0	2	0	0
Emergency Stop Buttons	2	2	1	1	1	2	1	2	0
Touch Display	0	1	0	1	1	2	1	1	0

For a reliable development process leading to a successful robot system application, an adequate development method must be applied, which guarantees that all the requirements of the specific application are followed in all technological decisions.

This method must be adapted to the complex design processes of service robots. The definition of user and system requirements must be arranged from the viewpoint of users, operators, and market needs. The customers of the service robots should be included in the creation process of these documents. The technical decisions during the development process might be supported by decision engineering methods to weight the given user and system requirements as well as the experiences from previous system developments in order to select appropriate technologies.

For the design of the two robot platforms developed in the frame of this book, the V-Model was applied as the systematic development method, which is described in the following chapter. The compilation of user and system requirements was arranged in cooperation with users and operators, involved in the projects. For the determination of technical solutions, the knowledge-based decision method AHP [Saaty, 1994] was applied (Chapter 4).

Chapter 3

System Design Models

This chapter describes design models for system development in the field of embedded hard- and software systems including an overview of the Waterfall-Model, the Spiral-Model, the Prototyping Model, and the Iterative-Incremental Model. The V-Model, as the methodical approach used in this work, is introduced in more detail. The adaptation of this approach for the development of service robot platforms is discussed.

3.1 Common Concepts of System Design Models

Design models are used to structure the development of complex systems. The common idea is the breakdown of the development process into subtasks that can be executed separately. Standardized design methods provide powerful strategies, technologies, and tools to complete the subtasks and to ensure an effective and determined design process. They simplify the assignment of time, manpower, and budget to a subtask. Every subtask receives defined inputs from previous tasks and generates outputs to following tasks.

The life cycle of every technical system is described by the System Development Life Cycle (SDLC). It is the meta-model of every design model and describes the stages of a system development and application period of a product (Figure 3.1).

Every system development process begins with the initiation phase. During this period, the purpose for a new system is documented and the high-level requirements are defined. The following development phase handles the system breakdown into sub-modules, the realization of these modules, its integration into the system, and initial system testing. Most of the design models focus on the development phase because of the complexity of the different tasks within this phase. During the following implementation period, the system is installed in the final application. Minor improvements are carried out during the operation and maintenance period. Over time, the system becomes obsolete because of the general technical progress. The performance of the system components becomes limited or is no longer compatible to novel technologies. This leads to the disposal period. The SDLC restarts with the initiation phase in which again a new specification of system operation parameters is required.

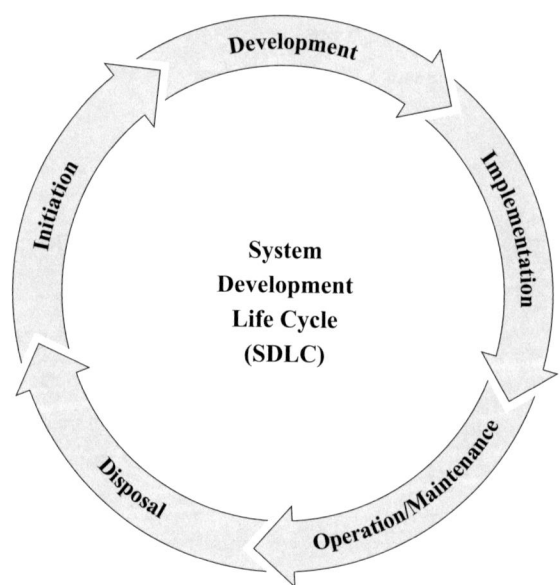

Figure 3.1: *Overview of the System Development Life Cycle (SDLC) [Radack, 2009].*

3.2   The Waterfall-Model

The Waterfall-Model was the first structured approach for software system developments, introduced by Royce in 1970. The first version of the model contained seven phases: system requirements, software requirements, analysis, program design, coding, testing, and operation. A later version of the model, developed by Versteegen, consisted of just four phases: requirements analysis, design, implementation, and integration [Versteegen, 2002]. The number of phases in the Waterfall-Model varies for different model implementations. A higher number of phases results in a more detailed decomposition of the design process into subtasks. A generally accepted version of the Waterfall-Model consists of eight phases (Figure 3.2), which is also used in non-software development projects [Heinrich, 2007].

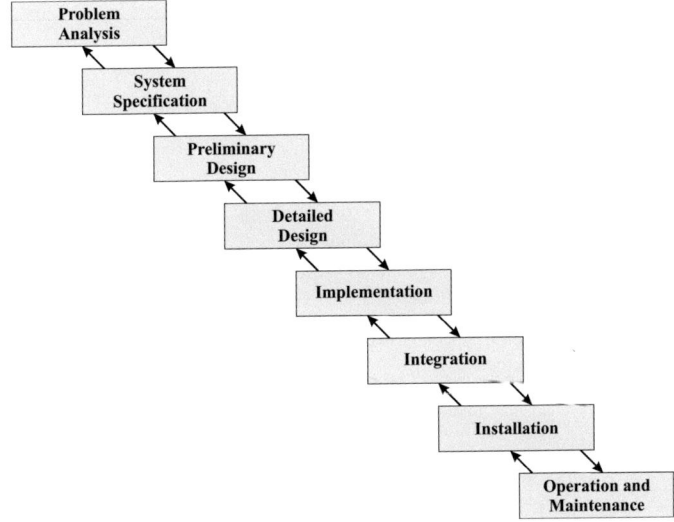

Figure 3.2: *The Waterfall-Model based on eighth phases [Heinrich, 2007].*

The Waterfall-Model consists of development phases which are sequentially executed. One phase transits to the following when the predecessor phase has generated the transfer objects. The completion of one phase can be seen as a milestone within the process. Novel versions of the Waterfall-Model also include recurrent transitions of later phases back to previous. The linear layout of this model simplifies the planning of costs and timing of each phase.

The constraint of the Waterfall-Model is that linear transitions between phases require high effort if a misconception of a previous step is noticed in a late project phase. In this case, the model has to be retraced step by step to the phase where the failure can be solved. Another limitation is the quality management that is restricted to each single phase. An overall quality management system is not available in the Waterfall-Model. A third disadvantage is the compilation of the system specification at the beginning of the development. An adaptation or extension of the system specification depending on development results in later stages is not provided in this model. Finally, the production of intermediate products for evaluation by the customer is not intended. This late inclusion of the customer increases design failures.

3.3 The Spiral-Model

The Spiral-Model is a further improvement of the Waterfall-Model and orients strictly on risk minimization during the development process [Boehm, 1988]. This is addressed by the creation of four phases that are continuously repeated until either the system development is finished or the project development has failed. The first phase covers the definition of project goals and possible alternatives, and the determination of constraints. The following phase evaluates alternatives searching for risk minimizing solutions. The third phase is the development and test period. The fourth phase prepares the next entrance into phase one.

The name of the Spiral-Model is derived from the spiral form of the development process. The project development starts at the center point and evolves passing the four model phases characterized by two axes: cumulative costs and review by the customer (Figure 3.3). The angle of the spiral function describes the current project state within a cycle and the area covered by the function represents the expenses of the project. The number of cycles shows the progress of the project.

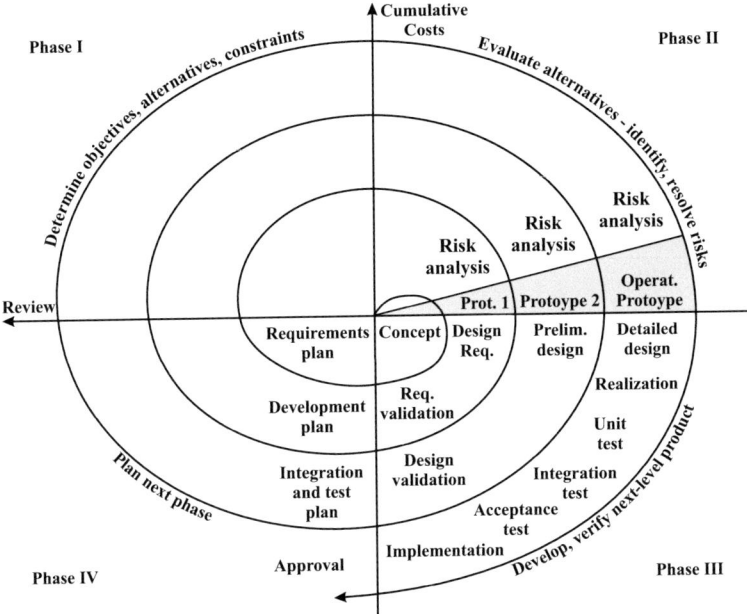

Figure 3.3: *The Spiral-Model [Boehm, 1988].*

The advantage of the Spiral-Model to the Waterfall-Model is the involvement of the customer in every cycle. The customer is required to accept the current version of the prototype. Technical and functional design errors can be identified at an early stage. If a development phase fails, the prototype of the last cycle is used to evaluate alternative solutions. The combination of development and maintenance in the Spiral-Model is a good approach for long-term design strategies.

3.4 The Prototyping and the Iterative-Incremental Model

In contrast to the Waterfall-Model and the Spiral-Model, the Prototyping Model does not belong to the group of phase models. The idea of the Prototyping Model is the creation of preliminary systems for evaluation and testing at an early project stage. The generated prototypes are not meant to cover the complete requirements of the final system specification. They are mainly used to highlight selected aspects or to present technical solutions to customers for an early feedback [Smith, 1991].

A special method of prototyping is *Evolutionary Prototyping*, in which the prototypes are recycled for further developments and adaptations. Therefore, this development process is not determined and the distinction between development and maintenance is usually not given. An example for the usage of *Evolutionary Prototyping* in the field of mobile robots is the development of an American Association for Artificial Intelligence (AAAI) competition robot by the Université de Sherbrooke, in Québec, Canada [Michaud et al., 2009]. In 2011, the third version of this mobile robot is developed that will participate at the next AAAI robot competition. So far, every predecessor version (first version attended to the competition in 2000, second version in 2005 and 2006) formed the basis for the development of the next system.

Similarly, the Iterative-Incremental Model generates prototypes to gather properties and parameters of the system [Heinrich, 2007]. In contrast to the Prototyping Model, this approach starts with a well-structured system that is gradually expanded by new functionalities during development. This allows for the detection of design mistakes at an early project stage to save costs and time. However, the adaptation of the prototype to the increasing demands requires a lot of effort.

3.5   The V-Model

The V-Model is derived from the Waterfall-Model and was first published by the German Federal Ministry of Defense (BMVg) in 1986. Further improvements were the V-Model 92 (1992) and the V-Model 97 (1997), which contained first iterative elements to allow for cyclic development processes [Heinrich, 2007]. Similar to the Waterfall-Mode, the V-Model defines phases that have to be carried out during the project. The composition of these phases depends on the content of the project and the stakeholders. For example, a system development project between a supplier and an acquirer would require phases like *Offer Submitted* and *Contract Awarded* whereas a supplier project preparing system improvements does not need to include theses phases. The V-Model further supports iterations of development phases, if the quality of an output did not fulfill all requirements or if the requirements have changed during the project execution. The quality of the phase outputs determines the transition to the next development stage.

The current version of the V-Model is the V-Model XT (*eXtreme Tailoring*) [V-Model XT, 2009]. One special feature of this model is the *Tailoring*. This process enables the flexible adaptation of the V-Model XT to a variety of project types (e.g., software/hardware development projects, system integration projects, or control projects for contracted developments). *Tailoring* helps to determine development strategies matching exactly the demands of the project by including and excluding of development tasks.

The modularity of the project phases, the possibility of project stage iterations, and its flexible adaptation to different project types make this model suitable for complex development projects. This work applies the V-Model XT for the development of the robot platforms. The V-Model XT was used to define system requirements, to decompose both robot systems into units, to design and test the units, to join the units to systems, and to integrate and evaluate the completed robot systems.

The following sections introduce the V-Model XT (called V-Model in the following sections) focusing on aspects of a robot platform development process.

3.6 Adaptation of the V-Model to Service Robot Developments

3.6.1 Tailoring Process

The V-Model is applicable to a variety of project types, which explains the high complexity of this design model. The adaptation process of the V-Model to any particular development project, *Tailoring*, is carried out during the project and is one of the most important tasks during the execution of the V-Model.

Two types of *Tailoring* methods are provided by the V-Model: *Static Tailoring* and *Dynamic Tailoring*. *Static Tailoring* is applicable to projects with fixed development frameworks, e.g., design processes with a predefined product realization. It defines the project execution strategy and the required *Process Modules* (Section 3.6.3) before the start of the project. *Dynamic Tailoring* is used for non-predicable project frameworks, e.g., developments in which the concept might change during the project. It takes place in the course of the project and increases the flexibility of the design model. In this case, further *Process Modules* can be included during the development process or unnecessary modules can be removed.

The framework for the developments described by this book is well defined, because the stakeholders defined the expected project outputs at the beginning of the projects. Therefore, *Static Tailoring* can be applied.

3.6.2   Project Role, Project Type, and Project Type Variant

The *Tailoring* process starts with the determination of the *Project Role*, the *Project Type*, and an appropriate *Project Type Variant*. The *Project Role* is given by all stakeholders and can include the acquirer, the supplier, or both. It determines organizational tasks for the collaboration of the stakeholders. In the frame of this work, the *Project Role* includes the acquirer and the supplier (Figure 3.4). The acquirer (represented by the company Toom BauMarkt for the shopping robot development and nursing service provider for the home-care robot development) defines system requirements, provides the test bed, and evaluates the development results. Whereas the suppliers (represented by the scientific and technological partners) are responsible for the system design.

Possible *Project Types* are system development projects and projects for introduction and maintenance of organization-specific process models. The first type applies to this framework. In the next step, the subjects of the project have to be identified. Subjects can be: hardware systems, software systems, a combination of both, embedded systems, or system integrations. The design of a mobile service robot comprises software, hardware, and embedded systems.

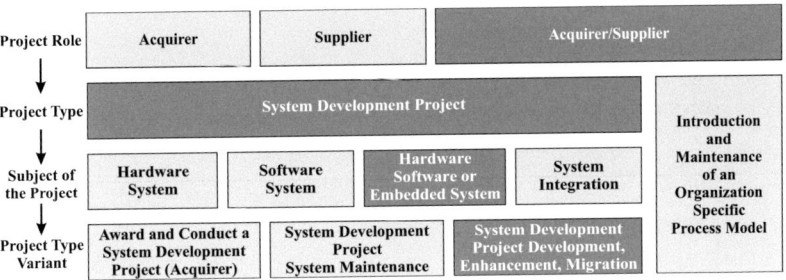

Figure 3.4: *Classification of the project execution strategy regarding the V-Model. The dark-gray highlighted project characteristics are relevant for the robot developments [V-Model XT, 2009].*

With the specification of the *Project Role*, the *Project Type*, and the subjects, the *Project Type Variant* can be identified. The *Project Type Variant* depends on the phase of the SDLC that is covered by the project (Figure 3.1). The developments of this work belong to the *Project Type Variant* for system development, enhancement, and migration.

3.6.3   Process Modules

The identified *Project Type Variant* specifies *Process Modules* that have to be included into the design process. *Process Modules* constitute the task-based view of the development process and involve self-contained units with defined inputs, specified tasks, and defined outputs. The composition of these modules determines the course of a development process based on the V-Model. Depending on the *Project Type Variant*, the V-Model defines *Core Process Modules* that are mandatory for every development process:

- Project Management,

- Quality Assurance,

- Configuration Management,

- Problem and Change Management,

- System Development,

- Specification and Requirements,

- Delivery and Acceptance (Supplier),

- Delivery and Acceptance (Acquirer).

In addition to mandatory *Process Modules*, optional modules might be integrated. The usage of optional modules depends on the subject of the project (Figure 3.4). For example, the module *Hardware Development* must be included as soon as a hardware unit is identified in the system architecture. Examples for optional *Process Modules* are:

- Hardware Development,

- Software Development,

- Evaluation of Off-the-Shelf Products,

- Usability and Ergonomics.

If a project includes the development of high-level software algorithms, human-robot interaction, or user application design, *Process Modules* like *Usability and Ergonomics* might be included in the development process. This goes beyond the focus of this work. A detailed description of available *Process Modules* including required inputs, outputs, and work tasks can be found in the V-Model documentation [V-Model XT, 2009].

3.6.4 Project Stages and Decision Gates

The structure-oriented view of the development process is based on *Project Stages* that have to be accomplished during the development, e.g., system specification or

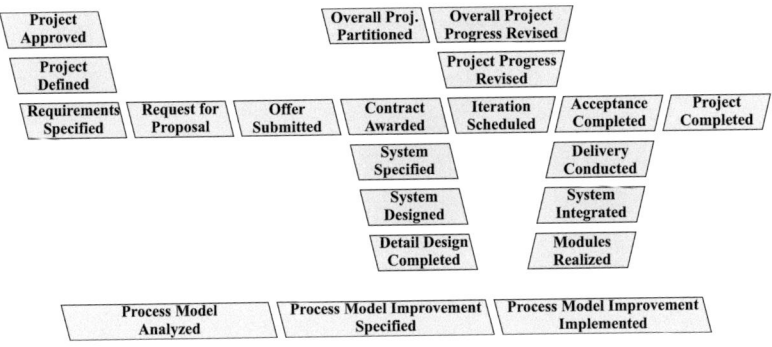

Figure 3.5: *Decision Gates defined by the V-Model representing milestones of the development process [V-Model XT, 2009].*

system design. Specific tasks of *Process Modules* are executed within these stages to generate required outputs. Depending on the project execution strategy, a specific combination and order of these stages is defined. The achievement of a goal of a *Project Stage* is marked by a *Decision Gate*, where the current status of the project has to be evaluated by the project management. Therefore, *Decision Gates* represent milestones of the development process (Figure 3.5). To pass a *Decision Gate* and to enter the next *Project Stage*, all tasks of the previous stage have to be finished. If the quality of one task outcome is insufficient, which prevents the passing of a *Decision Gate*, three solutions are considered by the V-Model: the task has to be revised until it has an appropriate quality; the project is traced back to permit an alternative solution or to repeat the processing of several predecessor task outputs; or the development project could be canceled.

The *Decision Gates* for the developments of the robot platforms are shown in Figure 3.6. The gates *Project Approved* and *Project Defined* were already passed before the development process started. These gates were completed by passing the funding agencies evaluation processes. The requirements specifications of the robot systems, which have to be done before the gate *Requirements Specified*, were included in the project proposals. Furthermore, most of the decision gates belonging to the acquirer-supplier interface were not included to the development process, because the related tasks were already carried out during the project application and evaluation pro-

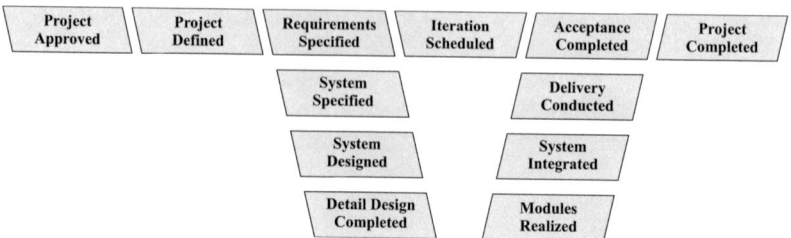

Figure 3.6: *Project-specific development process of the robot platforms.*

cesses. Therefore, the development processes of the robot systems described in this work started with the developments required for the decision gate *System Specified* and continued until the decision gate *Delivery Conducted* was passed. These *Decision Gates* are described in more detail in the following:

System Specified: The goal of this project stage is the overall system specification. The initial situation and the intended technical solutions have to be described. Functional requirements of the robot platform have to be derived from relevant use cases. Further, non-functional requirements have to be defined, e.g., the working area, physical constraints, or demands on service and installation. An overview of the system architecture should be included in the system specification and first technical approaches (like system interfaces) should be presented. For the later verification process of the outputs of this project stage, an evaluation specification based on test cases must be included. It is also recommended to specify a safety and security analysis of the developed system.

System Designed: This project stage includes the discussion of possible system architectures. After selecting the optimal architecture, the system decomposition must be carried out. Here, the system is broken down into subsystems (e.g., *Power Supply Subsystem*), segments (e.g., *Battery System Segment*), and units (e.g., *Battery Control Unit*). Elements identified on all hierarchical levels and the interfaces between these system elements should be described. The requirements of subsystems and segments must be specified. All design decisions at these levels must be comprehensively documented. Further, an evaluation specification for each system element should be prepared.

Detail Design Completed: The tasks within this project stage deal with the design of system units. The elements at this hierarchical level must be assigned to one of the groups: hardware units exclusively containing hardware components; software units exclusively build from software components; embedded system units containing hardware and software components; or external units supplied by third parties. For every unit, the hardware and software architecture must be specified. Every unit is further decomposed into smaller components. An overview of the internal interfaces between those unit components and the corresponding exchanged information should also be included. At the end of this process, an evaluation specification must be prepared for the verification of the functionality of every unit.

Modules Realized: In this design stage, all identified units have to be realized physically. Based on the specification of the previous project stage, the design of the units has to consider the described software and hardware architectures, required components, and intended interfaces. The realized modules and the functional testing have to be documented.

System Integrated: This decision gate requires the composition of all system elements. The developed units are combined to segments, segments to subsystems, and subsystems composed to the final robot system. All required evaluations that were defined in previous stages must be carried out. All processes have to be documented.

Delivery Conducted: In this final stage, the developed robot platform has to be tested regarding the evaluation specification defined in the decision gate *System Specified*.

After the successful fulfillment of the requirements of all previous decision gates, the project enters the stage, which leads to the decision gate *Acceptance Completed*. In this stage, user trials have to be carried out under real-life conditions to verify the functionality of the overall system. The development partners verify the functionality of the system in the intended operation area. The professional partners analyze the system regarding user acceptance, usability, and benefit. In case of a positive evaluation by all partners, the project can be finished by entering and passing the decision gate *Project Completed*. If the quality of the developed system does not fulfill all requirements (e.g., because of changed system requirements), another development iteration can be initialized. Usually, the decision gate *Iteration Scheduled* re-enters the project stage leading to the decision gate *Requirements Specification*. If the system requirements do not have to be adapted, the iteration can start directly with the revision of the overall system specification or the system design process.

For the design of the shopping robot platform, three development cycles were scheduled during the *Tailoring* phase of the project (Figure 3.7). The first cycle was arranged for the design of an early prototype. This system included only the necessary components for first software implementations and functional analysis of system elements. Because of its prototype character, this robot was not evaluated based on the entire system specification (project stage *Delivery Conducted*). The second design iteration included the main development process of the shopping robot platform. In this cycle, all required system elements were developed and verified. The output was a fully functional robot system (*SCITOS G5*) that was used for the

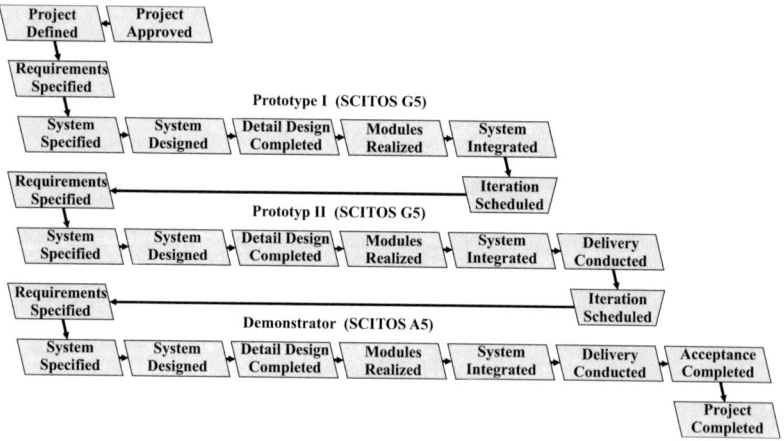

Figure 3.7: *System development process on the example of the shopping robot platform carried out in three iterations. The design of the home-care robot system was carried out in two iterations leaving out the first prototype stage.*

implementation of all system functionalities and first user trials. The third design cycle was scheduled to make minor changes on the system to optimize it for the usage in public environments. This primarily included changes on the casing of the robot.

The design process of the home-care robot was planned to be carried out in two development cycles. This was reasonable, because of the gained knowledge of the research partners from the shopping robot project that could be applied to this development. Furthermore, the shopping robot platform could be used for first software implementations and tests. The novel home-care robot platform, designed in the first development cycle, could already be validated regarding the specified evaluation requirements and applied for user trials. The second cycle was executed to make minor changes on the robot platform based on information collected from the first development cycle.

The successful application of the V-Model requires design decisions at different abstraction levels (from system level down to unit level). In the next chapter, the Analytic Hierarchy Process (AHP) will be described. This decision method was used in this work to support the V-Model based development processes.

Chapter 4

The Analytic Hierarchy Process for Decision-Making

This chapter describes the Analytic Hierarchy Process (AHP) as a decision engineering method to solve complex multi-criteria problems. The AHP is used in several areas like industry, business, or education to support complex decision processes. In the field of robotics, the AHP has been applied to the selection process of available robot systems (e.g., [Athawale and Chakraborty, 2011] and [Özgürler et al., 2011]) and the optimization of control strategies for robot manipulators (e.g., [Banga et al., 2007]). There seems to be no (documented) application of the AHP to the development process of service robots.

This work introduces and adapts the AHP as a decision method for the design of complex mobile robot systems. In the first section of this chapter, the principles of the AHP are described. The detailedness of this description should allow the better traceability of later calculations. The next section generates the decision problems and the AHP structures for the service robot developments. In Section 4.2.1, criteria applicable to service robot developments are discussed. The following sections present design alternatives for service robots and weight these alternatives under the consideration of the defined criteria.

4.1 Principles of the Analytic Hierarchy Process

The AHP was developed by the mathematician Thomas L. Saaty in the 1970s to facilitate complex design decisions [Saaty, 1977]. Multiple criteria have to be considered in complex decisions. The AHP applies pairwise comparisons between all decision criteria to derive a prioritization of the criteria (weights). It also supports the decomposition of criteria into sub-criteria for highly complex decision processes or ambiguous criteria [Saaty, 1994]. Decision-making processes based on the AHP can be used to select the best solution (e.g., means of travel) out of given alternatives (e.g., car, train, airplane, bus). These alternatives are also weighted based on pairwise comparisons. The evaluation of alternatives must be carried out under consideration of each criterion (e.g., time, comfort, costs). The resulting weights of the alternatives for each criterion and the weights of the criteria are used to calculate the overall priorities of the alternatives. These priorities represent the suitability of the alternatives to solve the given decision problem.

During the execution of the AHP, the decision hierarchy is constructed, the criteria are weighted, the alternatives are compared, and the overall priorities are determined. In the next sections, the sequence and the mathematics of the AHP are introduced based on an example decision problem.

4.1.1 Construction of the AHP Hierarchy

At the beginning, the problem has to be defined. This is particularly important, when several parties weight the criteria. A common goal for the problem solution is required to get consistent weights of the criteria. Alternatives and criteria have to be defined that are relevant for the decision process. Both elements determine the structure and complexity of the AHP hierarchy. Alternatives should be proved regarding mandatory requirements. Alternatives that do not fulfill all mandatory requirements should be removed from the decision process at this early stage to simplify the pairwise comparison process.

Based on the determined criteria (N) and the alternatives (M), two decision matrices can be generated. The first matrix consists of the weights a_j of the criteria:

$$A = \begin{bmatrix} a_1 & ... & a_j & ... & a_N \end{bmatrix} \qquad (4.1)$$

The second matrix contains the weights b_{ij} of alternatives i for each criterion j:

$$B = \begin{bmatrix} b_{11} & ... & b_{1j} & ... & b_{1N} \\ ... & & ... & & ... \\ b_{i1} & ... & b_{ij} & ... & b_{iN} \\ ... & & ... & & ... \\ b_{M1} & ... & b_{Mj} & ... & b_{MN} \end{bmatrix} \qquad (4.2)$$

Figure 4.1 presents the AHP hierarchy for the example of the problem to find the best means of travel from one city to another. In the resulting AHP hierarchy, the decision matrices have the dimensions $dim(A) = 3$ (because of three considered criteria) and $dim(B) = (4 \quad 3)$ containing the weights of the four alternatives for each criterion.

4.1.2 Definition and Weighting of Criteria

Once the hierarchy and the decision matrices have been generated, the AHP uses pairwise comparisons of criteria to assess their impact on the decision goal. The resulting comparisons (Table 4.1) reveal the importance of one criterion relative to another. All comparisons e_{ij} of a criterion i evaluated with criterion j are represented in the evaluation matrix E:

$$E = \begin{bmatrix} e_{11} & ... & e_{1j} & ... & e_{1N} \\ ... & & ... & & ... \\ e_{i1} & ... & e_{ij} & ... & e_{iN} \\ ... & & ... & & ... \\ e_{N1} & ... & e_{Nj} & ... & e_{NN} \end{bmatrix} \qquad \begin{array}{l} \forall i = j : \quad e_{ij} = 1 \\[6pt] \forall i = 1, ..., N \quad \forall j = 1, ..., N : \quad e_{ij} > 0 \\[6pt] \forall i = 1, ..., N \quad \forall j = 1, ..., N : \quad e_{ij} = e_{ji}^{-1} \end{array} \qquad (4.3)$$

Table 4.1: *Absolute numbers for the weighting of comparisons [Saaty, 1980].*

Intensity of Importance	Definition	Explanation
1	Equal importance	Two activities contribute equally to the objective.
3	Weak importance of one over another	Experience and judgment slightly favor one activity over another.
5	Essential or strong importance	Experience and judgment strongly favor one activity over another.
7	Demonstrated importance	An activity is strongly favored and its dominance demonstrated in practice.
9	Absolute importance	The evidence favoring one activity over another is of the highest possible order of affirmation.
2, 4, 6, 8	Intermediate values between the two adjacent judgments	When compromise is needed.
Reciprocals of above nonzero	If activity i has one of the above nonzero numbers assigned to it when compared with activity j, then j has the reciprocal value when compared with i.	

A drawback of this approach is that inconsistency within the evaluation matrix might arise. One of the reasons is that absolute numbers are used as judgments, which do not perfectly reflect the relations between all criteria. An unacceptable inconsistency might occur, when logical mistakes are made during the comparison process. An example would be the comparison of three criteria: A is more important than B, B is more important than C, and C is more important than A. For the estimation of the inconsistency of an evaluation matrix, the *Consistency Index C.I.* and the *Consistency Ratio C.R.* can be calculated. Both values use the fact that the maximal eigenvalue λ_{max} of a consistent evaluation matrix (E) is equal to the matrix dimension $N = dim(E)$. An increasing inconsistency leads to an increasing λ_{max}. The consistency values $C.I.$ and $C.R.$ are defined as:

$$C.I. = \frac{\lambda_{max} - N}{N - 1} \tag{4.4}$$

$$C.R. = \frac{C.I.}{R.I.} \tag{4.5}$$

The *Random Index R.I.* is the mean *Consistency Index* of randomly generated matrices that follow the attributes of equation 4.3 with a matrix dimension greater than two. For smaller matrix dimensions this value is defined as zero (Table 4.2).

Table 4.2: *Random Index R.I. [Saaty, 1980]*

N	1	2	3	4	5	6	7	8	9	10
R.I.	0.00	0.00	0.52	0.89	1.11	1.25	1.35	1.40	1.45	1.49

The cutoff value, which requires a revision of the evaluation matrix is $C.R. > 0.1$.

After the generation of the evaluation matrix and the verification of the matrix consistency, the criteria weights can be calculated. Therefore, the evaluation matrix has to be normalized by column sums:

$$D = \begin{bmatrix} d_{11} & \cdots & d_{1j} & \cdots & d_{1N} \\ \cdots & & \cdots & & \cdots \\ d_{i1} & \cdots & d_{ij} & \cdots & d_{iN} \\ \cdots & & \cdots & & \cdots \\ d_{N1} & \cdots & d_{Nj} & \cdots & d_{NN} \end{bmatrix} = \begin{bmatrix} \frac{e_{11}}{\sum\limits_{i=1}^{N} e_{i1}} & \cdots & \frac{e_{1j}}{\sum\limits_{i=1}^{N} e_{ij}} & \cdots & \frac{e_{1N}}{\sum\limits_{i=1}^{N} e_{iN}} \\ \cdots & & \cdots & & \cdots \\ \frac{e_{i1}}{\sum\limits_{i=1}^{N} e_{i1}} & \cdots & \frac{e_{ij}}{\sum\limits_{i=1}^{N} e_{ij}} & \cdots & \frac{e_{iN}}{\sum\limits_{i=1}^{N} e_{iN}} \\ \cdots & & \cdots & & \cdots \\ \frac{e_{N1}}{\sum\limits_{i=1}^{N} e_{i1}} & \cdots & \frac{e_{Nj}}{\sum\limits_{i=1}^{N} e_{ij}} & \cdots & \frac{e_{NN}}{\sum\limits_{i=1}^{N} e_{iN}} \end{bmatrix} \tag{4.6}$$

Afterwards, the row sums of this normalized evaluation matrix D are calculated and divided by the dimension N of the matrix. These values represent the weights a_i for the criteria (refer to equation 4.1):

47

$$a_i = \frac{\sum\limits_{j=1}^{N} d_{ij}}{N} \quad \forall i = 1, ..., N \tag{4.7}$$

The described evaluation procedure based on pairwise comparisons is applicable for qualitative information. If quantitative information about the criteria is available, a different mathematical approach can be used [Meixner and Rainer, 2002]. The advantage of this approach is that it does not create inconsistency. The weights of the criteria a_i can be calculated based on the given numerical values v_i:

$$a_i = \frac{v_i}{\sum\limits_{j=1}^{N} v_j} \quad \forall i = 1, ..., N \tag{4.8}$$

Based on this equation, a higher value of v_i generates a higher weighting value a_i. If a numerical value of a criterion has to be minimized in the decision process (e.g., system costs), the following calculation applies:

$$a_i = \frac{\frac{1}{v_i}}{\sum\limits_{j=1}^{N} \frac{1}{v_j}} \quad \forall i = 1, ..., N \tag{4.9}$$

For the given example (Figure 4.1), three criteria were defined: time (C_1), comfort (C_2), and costs (C_3). Assuming that the example refers to a business trip, the following statements could be derived relying on the evaluations of Table 4.1: Time is weakly more important than comfort (3/1), time is "demonstrated" more important than costs (7/1), and comfort is essentially more important than costs (5/1). The resulting evaluation matrix is:

$$E_{travel} = \begin{bmatrix} 1/1 & 3/1 & 7/1 \\ 1/3 & 1/1 & 5/1 \\ 1/7 & 1/5 & 1/1 \end{bmatrix} \tag{4.10}$$

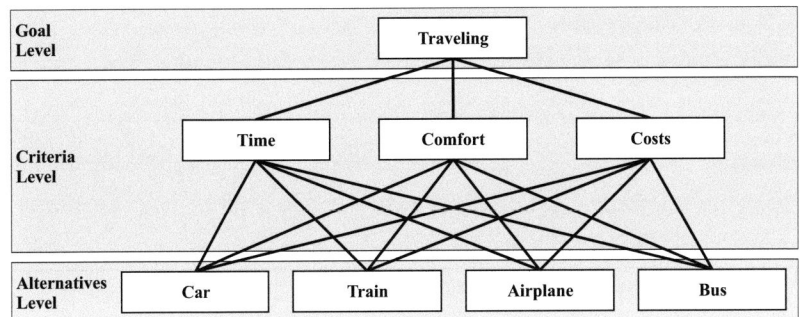

Figure 4.1: *Example AHP hierarchy consisting of the three abstraction levels for the decision goal, criteria, and alternatives.*

To prove the consistency of the evaluation matrix, the corresponding *Consistency Ratio C.R.* for a three-dimensional matrix can be determined:

$$C.R._{travel} \approx \frac{0.032}{0.52} \approx 0.062 \leq 0.1 \tag{4.11}$$

Based on equations 4.6 and 4.7, the criteria weights can be calculated:

$$a_1 \approx 0.643 \qquad\qquad (Time) \tag{4.12}$$
$$a_2 \approx 0.283 \qquad\qquad (Comfort) \tag{4.13}$$
$$a_3 \approx 0.074 \qquad\qquad (Costs) \tag{4.14}$$

The resulting weights matrix of equation 4.1 is

$$A_{travel} \approx \begin{bmatrix} 0.643 & 0.283 & 0.074 \end{bmatrix} \tag{4.15}$$

Consequently, the traveling time is the criterion with the highest priority (64.3 %) for this decision process. The comfort is less important (28.3 %) and the traveling costs have the lowest priority (7.4 %).

4.1.3 Comparison of Alternatives

The weighting process of the alternatives follows the same rules as the weighting process of the criteria. The pairwise comparison method has to be applied for qualitative information; the numerical method applies if quantitative information are available. The weighting process of the alternatives must be carried out under consideration of every criterion separately.

Considering the given example, four alternatives were defined: a car (A_1), train (A_2), airplane (A_3), and a bus (A_4). Therefore, the evaluation matrices have the dimensions $M = 4$. The comparison process of these alternatives considering the traveling time may have the following results (compare Table 4.1): The car needs the same time as the train (1/1), essentially more time than the airplane (1/5), and "demonstrated" less time than the bus (7/1). The train needs essentially more time than the airplane (1/5) and "demonstrated" less time than the bus (7/1). The airplane needs absolutely less time than the bus (9/1). The resulting evaluation matrix under consideration of the traveling time is:

$$
E_{time} = \begin{bmatrix} 1/1 & 1/1 & 1/5 & 7/1 \\ 1/1 & 1/1 & 1/5 & 7/1 \\ 5/1 & 5/1 & 1/1 & 9/1 \\ 1/7 & 1/7 & 1/9 & 1/1 \end{bmatrix} \tag{4.16}
$$

The comparison process regarding the comfort generates the statements: The car is weakly less comfortable than the train (1/3), essentially less comfortable then the airplane (1/5), and weakly more comfortable than the bus (3/1). The train is weakly less comfortable than the airplane (1/3) and essentially more comfortable than the bus (5/1). The airplane is "demonstrated" more comfortable than the bus (7/1).

$$E_{comfort} = \begin{bmatrix} 1/1 & 1/3 & 1/5 & 3/1 \\ 3/1 & 1/1 & 1/3 & 5/1 \\ 5/1 & 3/1 & 1/1 & 7/1 \\ 1/3 & 1/5 & 1/7 & 1/1 \end{bmatrix} \tag{4.17}$$

The *Consistency Ratios* C.R. for both matrices are:

$$C.R._{time} \approx \frac{0.080}{0.89} \approx 0.090 \leq 0.1 \tag{4.18}$$

$$C.R._{comfort} \approx \frac{0.039}{0.89} \approx 0.044 \leq 0.1 \tag{4.19}$$

The decision matrix for this example including the first two comparison processes is:

$$B_{travel} \approx \begin{bmatrix} 0.176 & 0.122 & b_{13} \\ 0.176 & 0.263 & b_{23} \\ 0.609 & 0.558 & b_{33} \\ 0.039 & 0.057 & b_{43} \end{bmatrix} \tag{4.20}$$

The analysis of this matrix shows that the criterion *Time* (column one) is best achieved by the alternative *Airplane* (row three) with a weight of 60.9 %. The same alternative also complies best with the criterion *Comfort* (column two): 55.8 %.

The third criterion (*Costs*) can be evaluated based on the numerical value of the traveling prize. For this example, the following costs are assumed:

$$\begin{aligned} v_1 &= 100 & (Car) \\ v_2 &= 200 & (Train) \\ v_3 &= 1000 & (Airplane) \\ v_4 &= 25 & (Bus) \end{aligned} \tag{4.21}$$

The calculation of the weights is based on Equation 4.9, because lower costs are preferred in this decision. For this criterion, the alternative *Bus* gets the highest priority with 71.4 %, which is shown in the completed weights matrix (column three):

$$B_{travel} \approx \begin{bmatrix} 0.176 & 0.122 & 0.179 \\ 0.176 & 0.263 & 0.089 \\ 0.609 & 0.558 & 0.018 \\ 0.039 & 0.057 & 0.714 \end{bmatrix} \tag{4.22}$$

4.1.4 Determination of the Overall Priority

The final step is the determination of the overall priorities P based on the weights matrix A of the criteria and the weights matrix B of the alternatives:

$$P = A \cdot B^T = \begin{bmatrix} a_1 & ... & a_j & ... & a_N \end{bmatrix} \cdot \begin{bmatrix} b_{11} & ... & b_{i1} & ... & b_{M1} \\ ... & & ... & & ... \\ b_{1j} & ... & b_{ij} & ... & b_{Mj} \\ ... & & ... & & ... \\ b_{1N} & ... & b_{iN} & ... & b_{MN} \end{bmatrix} \tag{4.23}$$

For the given example:

$$P_{travel} \approx \begin{bmatrix} 0.643 & 0.283 & 0.074 \end{bmatrix} \cdot \begin{bmatrix} 0.176 & 0.176 & 0.609 & 0.039 \\ 0.122 & 0.263 & 0.558 & 0.057 \\ 0.179 & 0.089 & 0.018 & 0.714 \end{bmatrix} \tag{4.24}$$

$$P_{travel} \approx \begin{bmatrix} 0.161 & 0.194 & 0.551 & 0.094 \end{bmatrix} \tag{4.25}$$

Consequently, the example decision problem is best solved by the alternative *Airplane* with an overall priority of 55.1 %, followed by the alternative *Train* with 19.4 %

and the *Car* with 16.1 %. The alternative *Bus* solves worst the decision problem with just 9.4 %.

4.2 Application of the AHP to Decision Problems of the Robot Developments

The AHP is applied in this work to the robot systems development processes to facilitate decisions concerning the design of the systems. The AHP helps the selection of technical solutions that fit the system requirements of the respective robot platform best. Based on the hierarchical structure of the V-Model dominating the system design process, design decisions at system, subsystem, or segment levels have a higher impact to the development process than decisions at unit or component levels (Figure 5.5). Therefore, the application of the AHP to design decisions at system, subsystem, and segment levels has the highest benefit for the design processes.

For interactive service robots, design engineers have to select appropriate technologies in the fields of control systems, sensors systems, drive technologies, power supply systems, and interaction systems (Figure 5.3). In this work, four design decision problems are selected for the application of the AHP: the decisions about the system architecture, the battery technology, the charging system, and the drive system. In this work, the AHP is not applied to the selection processes of sensor and interaction systems, because the compositions of these systems were given by software concepts developed in previous projects. The development of alternative concepts was not in the focus of this work.

4.2.1 Criteria Definition

For the evaluation process of the selected problems, this work defines seven criteria: adaptability, operation time, usability, robustness, safeness, features, and costs. These criteria are assumed to have a major relevance for the assessment of an interactive service robot. Additional criteria could be included in the decision process to further improve the quality of the decision outputs (e.g., maneuverability or maintainability). It might also be possible to define sub-criteria for the decision process (e.g., production costs and operational costs). However, for the developments of shopping and home-care robots, the combination of these seven criteria appears to be applicable.

The following paragraphs describe the applied criteria in more detail:

Adaptability (A): This criterion describes the flexibility of a system to be adapted to changed requirements or novel applications. Adaptability depends on available interfaces (e.g., mounting points, computer interfaces, system connectors) and the modularity of a system (e.g., distributed control nodes). It is also influenced by the system resources of a robot (e.g., computational power, battery capacity, payload). This criterion is satisfied by modular systems with ample system resources; robot systems highly specialized to their applications do not meet this criterion well.

Operating Time (O): The operating time is the period, during which a robot system is able to execute its dedicated tasks before it has to be recharged. This time depends on the power consumption of the system and the battery capacity. It is also influenced by the duration of the charging process, in which the usability of the robot is limited. A development process under this criterion should consider the integration of energy-saving components and functionalities to switch-off unused modules.

Usability (U): This criterion includes aspects of user-friendliness (primarily for end-users, but also for operators and service personal), acceptance by users, and the attraction of attention (for shopping robots). The usability depends on the appearance of a robot and the integration of interaction functionalities. Examples of technical components that influence usability from the hardware point of view are sensor systems (e.g., to detect a user), computing power (e.g., to estimate the position of a user), or the drive system (e.g., to move the robot with an adequate speed).

Robustness (R): This criterion describes the probability of the breakdown of a robot system, because of malfunctions of system components or damage of the system caused by persons or obstacles. The implementation of high-quality system components, a reasonable integration of the components, and the detection of fall outs (if possible, before they occur) improve the robustness. Redundancy of system components and a low system complexity further contribute to the robustness of a system.

Safeness (S): This criterion deals with hazards for persons and objects generated by the robot system. To increase the safeness security sensors could be integrated (e.g., to reliably stop all motors), a smooth casing could be designed without hard edges, or system components that produce high voltages or high temperatures could be avoided. Redundancy of system components also increases the safeness.

Features (F): In addition to required functionalities of a robot system, this criterion considers supplementary functionalities for possible future modifications of a system. Variety of features is important for robot platforms that might be used for further applications and system developments. Robot systems that satisfy this criterion provide ample system resources (e.g., computing power, battery capacity) for the integration of new functionalities, or novel sensor systems for a potential improvement of environmental perception (e.g., depth cameras).

Costs (C): This criterion considers production and operational costs. It is best fulfilled by low overall system costs. A system development process under this criterion must consider the life expectancy of a robot system. The integration of long-living components usually increases the production costs, but reduces the operational costs, because of less on-site services (e.g., different battery technologies).

The weighting process of the criteria is carried in the V-Model project stage *System Design*, where all system requirements are defined (Section 5.2 for the shopping robot platform, Section 6.2 for the home-care robot platform).

4.2.2 Decision Alternatives

System Architecture

An appropriate system architecture has to be chosen for a robot platform to meet all requirements. In the context of this work, system architecture means the composition of embedded systems control modules in combination with communication systems. The analysis of existing system architectures in the field of mobile robots, revealed four different approaches that are discussed in the following:

System Architecture A1: Centralized Control Unit

The first approach uses a centralized control architecture (Figure 4.2). The main control component, often an embedded PC, represents the only high-computational unit of the system. Peripheral modules, like motor controller, sensor systems, or human-machine interfaces are directly connected to the embedded PC. The communication to the peripheral modules is usually based on PC compatible interfaces, like USB or RS232. To extend available interfaces, additional hubs can be integrated.

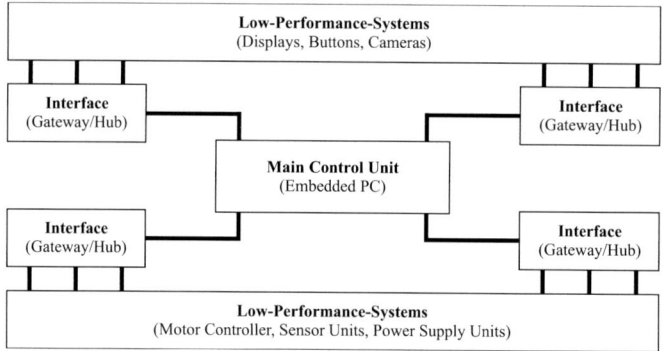

Figure 4.2: *System architecture A1 with centralized control unit.*

The advantage of this design architecture is the usage of commercial components leading to a cost-effective realization of a robot system. The disadvantages of this approach are the complex integration of system elements and the lack of redundancy, which make the robot system fault prone. Another disadvantage is that the processing of real-time-signals is not determinable by the embedded PC.

Centralized control architectures are, for example, implemented in the rover systems *K9* and *K10*, developed at the Intelligent Robotics Group at NASA Ames Research Center [Park et al., 2005]. These systems are equipped with laptop Personal Computers (PCs) that are connected to different hub modules. The variety of slave devices requires the connection of interfaces like RS232, RS485, USB, FireWire, or Ethernet to the internal PC. Another example is the robot system B21r built by RWI, which was initially used for the development of the shopping robot system. This robot is also equipped with a PC that is connected to the peripheral devices by an RS232 hub.

System Architecture A2: Centralized Control Unit with real-time Co-Controller

This system architecture addresses the challenge of processing real-time-signals by the integration of a real-time control unit cooperating with the integrated PC (Figure 4.3). This co-processing module is often realized by a Micro Controller (uC) or a Field Programmable Gate Array (FPGA) that is directly connected to all time critical system elements. Therefore, the system can give determined responses to time-critical events and can generate fast output signals, e.g., Pulse-Width Modulation (PWM) signals.

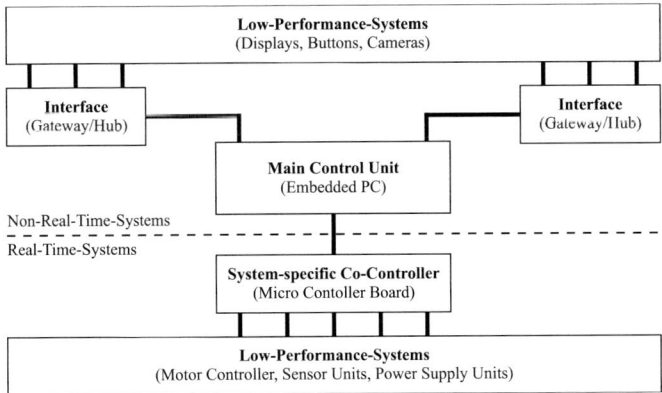

Figure 4.3: *System architecture A2 with centralized control unit and real-time co-controller.*

Compared to the first architecture, this version simplifies the realization of real-time processes. The implementation of a cooperation system based on a single board module is cost effective, especially at higher production quantities, where the development effort of this module is negligible. The disadvantage is the specialization of this module to the required tasks, which constricts the adaptability of this architecture. However, this architecture might be perfectly qualified for robot systems with high production quantities and an optimized cost-performance ratio.

One example is the robot system *MARVIN* (Mobile Autonomous Robotic Vehicle for Indoor Navigation) developed by the Mechatronics Group of the University of Waikato [Carnegie et al., 2004]. This system is equipped with a Windows PC and a hardware control unit, based on a powerful uC to avoid real time issues produces by the Windows operating system. A further example is the biped walking robot *Johnnie*, developed at the University of Technology, Munich [Lohmeier et al., 2004]. This system consists of two control PCs in combination with a PCI card with two uCs and powerful peripheral components for real-time tasks.

System Architecture A3: Decomposed high-performance Control Units

The third system architecture is based on distributed computation units (Figure 4.4). It consists of decomposed high-performance single board computers to control, e.g., interactive systems, motors, or sensor systems [Ahn et al., 2005]. Every component is connected to a PC that handles the incoming and outgoing information. The communication between different nodes is usually realized by high speed communication systems like Ethernet.

The advantage of this architecture is the distribution of system tasks to several powerful computation units. Depending on the complexity and the timing requirements of a task, an appropriate module can be chosen. This architecture further allows for the realization of redundancy to improve the system's reliability. The main disadvantages are the costs and the power consumption generated by the overhead of the implemented computation units.

For example, this architecture is implemented in the robot platform *EMIEW* developed by research laboratories of the company Hitachi [Hosoda et al.,]. This system

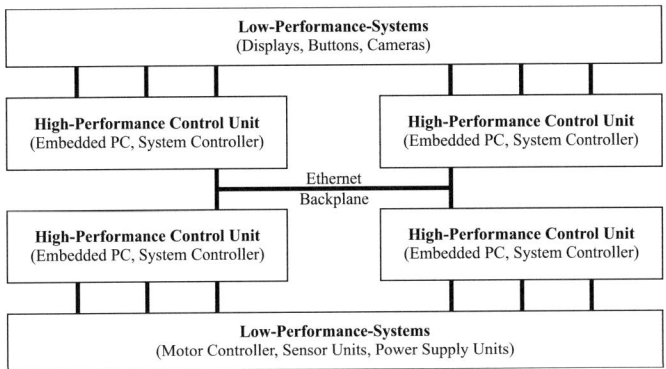

Figure 4.4: *System architecture A3 with decomposed high-performance control units.*

consists of four high speed control systems connected by an Ethernet network. Real-time tasks, like motor control or sensor analysis, are computed by real-time operating systems (installed on three control modules), whereas interaction is realized by one computational node with a non-real-time operating system. Another example of a professional robot system is the platform *MB835* by the company BlueBotics [Tomatis et al., 2004]. This system is equipped with two computation cards including an Intel PIII processor and a Motorola PowerPC system. The first system handles mobility tasks, the second system is responsible for interaction services. Further control modules are an I/O card, an encoder module, and a control module with a uC for security reasons.

System Architecture A4: Main Control Unit with decomposed Control Modules

The most flexible system architecture, similar to automotive architectures, is based on the distribution of the system to smart control modules, which are assigned to different tasks (Figure 4.5). Usually, these modules control real-time processes (e.g., motor or sensor control), whereas, complex tasks (e.g., localization or interaction) are still executed by non-real-time control units with high-computational power.

The advantage is the flexibility reached by the implementation of smart control

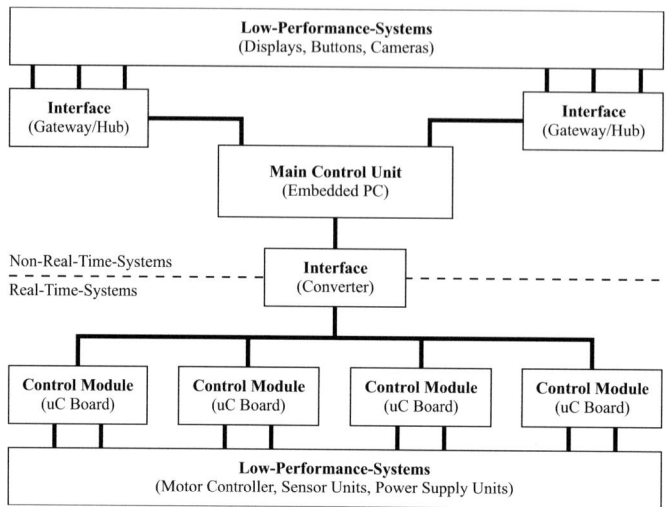

Figure 4.5: *System architecture A4 with main control unit and decomposed control modules.*

modules. These modules - optimized regarding complexity, costs, and power consumption - can be combined to adapt the robot's capabilities to the final application. The integration of a multi-master communication interface, e.g., Controller Area Network (CAN) or RS485 allows for a data exchange between different modules, which results in a fast and determined response to occurring events. The disadvantage are the higher production costs caused by redundancy in the module designs (e.g., each module is equipped with power supplies and communication interfaces).

Typically, mobile robot systems are based on this architecture. Examples are the interactive robots developed by the Fraunhofer Institute of Manufacturing, Engineering, and Automation (IPA) that use one embedded PC for high level computation, which is connected to low level control modules by a single CAN bus [Graf et al., 2004] and the modular mobile robot developed by the Beihang University Beijing using an RS485 bus to connect different low level control modules with a main control system [Zou et al., 2006]. Representatives with higher system complexities are the *HERMES* system developed by the Bundeswehr University Munich that uses a multi-processor main control system in combination with a single CAN bus [Bischoff and Graefe, 2004], the robot system *Kapeck* designed by the Universidade Federal do Rio Grande do Norte that uses a multi-embedded PC in combination with a single CAN bus for the interaction with real-time modules [Britto et al., 2008], or the robot platform *ARMAR-III* developed by the University of Karlsruhe consisting of five embedded PCs with a distributed communication system based on four CAN buses [Asfour et al., 2006].

Evaluation of System Architectures

The alternative system architectures are evaluated based on the previous defined criteria for the service robot development process (Section 4.2.1). The following statements and ratings can be derived:

Adaptability (A): The system architectures A1, A2, and A3 allow for a comparable adaptation to new requirements, because additional components can be connected to available interfaces of the embedded PC(s). Notably, system architecture A4 is evaluated as essentially more adaptable to new functionalities, because it additionally provides a highly flexible low-level communication bus for the connection of additional modules.

Operation Time (O): The system components with the highest power consumptions are embedded PCs. Other modules, like hubs or low-level control modules consume significant less energy. Therefore, architectures A1 and A2 are expected to consume a similar amount of energy; architecture A4 slightly more than A1 and A2. The power consumption of architecture A3 is expected to be considerably higher, because of the multiple PCs systems.

Usability (U): All careful designed system architectures should fulfill this criterion given the system requirements. Therefore, this criterion does not apply to the evaluation process of the system architecture.

Robustness (R): For the evaluation of this criterion, failures of hardware components and internal communication systems are not considered, because malfunctions of these systems are (almost) avoidable by reliable design and production processes. Therefore, robustness of a system architecture is primarily influenced by the complexity of the embedded software. A partitioning of software tasks among control modules would, therefore, improve the robustness of a system. Architecture A1 is expected to provide the lowest robustness, because all software functionalities are executed on one single computational unit. Architectures A2 and A3 provide a higher robustness, because of the distribution of tasks to several control units. The modular concept of architecture A4 allows for a flexible allocation of software functionalities to several responsible modules, which further increases the system robustness.

Safeness (S): Similar to robustness, this criterion is also influenced by the distribution of tasks to different modules. This allows for redundancy in the system (A3 and A4). Another aspect is the system reaction time to critical events, which is best satisfied by real-time systems (A2 and A4). For the criterion *Safeness*, a determinable reaction time is more important than redundancy, because mobile systems especially require fast and reliable reactions to critical situations (e.g., collisions).

Features (F): The available computing power of a system architecture to execute new software algorithms has to considered. This criterion is equally satisfied by system architectures A1 and A2, because both architectures provide a similar computing power. The architecture A3 is evaluated as essentially more important than A1 and A2, because the of integration of several PCs. The system architecture A4 is weighed to be considerably more important than A1 and A2, because of the integration of multiple control modules running embedded software.

Costs (C): The costs for a system architecture depend on the integrated components. Therefore, A3 obtains the worst rating, because PCs are the most expensive components of a system architecture. The redundancy of architecture A4 also produces additional costs for electrical components, power supplies, and casings. The most cost effective architectures are A1 and A2.

The pairwise comparison results are presented in Appendix A.1. The derived weights are summarized in Table 4.3.

Table 4.3: *Comparison results for system architectures.*

	Characteristic	A	O	U	R	S	F	C
A1	Central	12.5%	36.4%	25.0%	5.4%	5.7%	8.3%	43.4%
A2	Co-Controller	12.5%	36.4%	25.0%	14.6%	26.3%	8.3%	43.4%
A3	Multiple PCs	12.5%	6.6%	25.0%	23.7%	12.2%	41.7%	4.0%
A4	Modular	62.5%	20.7%	25.0%	56.3%	55.8%	41.7%	9.2%

Battery Technology

The battery technology of a mobile platform is critical for the availability of the system, and is an important factor for the production and service costs. Relevant parameters for the decision about a battery technology are the energy density, the maximum charging energy, costs, and the lifetime. The energy density defines the amount of energy that can be stored in a given volume. A higher energy density allows for longer operation times without recharging. Additionally, the maximum charging power influences the operation time of the robot, because it defines the idle state time of a robot during recharging. The parameters costs and lifetime influence the overall costs of a robot system. In the case that the lifetime of the battery is shorter than the lifetime of the robot system, production and service costs generated by the battery system have to considered in the evaluation.

Currently, four battery technologies are state-of-the-art and applicable to mobile service robots: lead-acid batteries, Nickel-Metal Hydride (Ni-MH) batteries, and lithium batteries based on $LiCoO_2$ and $LiFePO_4$. Examples of robot systems based on Lead-Acid batteries are the robot platforms *B21r* of the company RWI or the *Pioneer* robots of the company MobileRobots. Ni-MH batteries were, for instance, integrated into the service robot *Nanisha*, developed by the Universidad Popular Autónoma del Estado de Puebla in Mexico [Vargas et al., 2009]. The number of robot systems using lithium batteries is increasing, because this technology provides promising technical parameters. Examples are the service robots *REEM-H2* (Section 2.1.5), the *Care-O-Bot 3* (Section 2.2.2), or the mobile platform *Rollin Justin* by the German Aerospace Center [Fuchs et al., 2009]. Other energy storage technologies (e.g, capacitors or fuel-cells) could also be considered for the evaluation process, which goes beyond the focus of this work.

The evaluation process is carried out on specific example types of the four battery technologies. Even if the evaluation process does not depend on the battery configuration, the specific battery parameters are presented on cell configurations, applicable to both robot applications. It is assumed that the battery for the shopping robot and the home-care robot has a nominal voltage of about 24 V. For the integration of the battery inside the robot's chassis, a volume of about 12 dm^3 should be provided.

Battery Technology B1: Lead-Acid battery LC-X1242AP

This battery type consists of six series-connected sub-cells (each with a voltage of 2.0 V) resulting in a nominal cell voltage of 12.0 V. The nominal capacity is 42.0 Ah [Panasonic, 2011]. For the evaluation process, two of these cells connected in series are considered.

Battery Technology B2: Ni-MH battery D9000MAH-FT-1Z

The selected battery cell has a nominal voltage of 1.2 V and a capacity of 9.0 Ah [Emmerich, 2011]. A configuration of 20 cells in series and three cells in parallel could be integrated into the given space. The resulting battery, assembled from 60 cells, has a nominal voltage of 24.0 V and a capacity of 27.0 Ah.

Battery Technology B3: LiCoO$_2$ battery LP9675135

This cell type belongs to the group of lithium-polymer cells. It has a nominal voltage of 3.7 V and a capacity of 10.0 Ah [Dynamis Batterien, 2009]. The evaluated battery pack is composed by a matrix of seven cells in series and eight cells in parallel. The theoretical volume of such a battery is 6.3 dm^3. In practice, such cells require additional space for safety precautions. The 56 cells provide a nominal voltage of 25.9 V and a capacity of 80.0 Ah.

Battery Technology B4: LiFePO₄ battery GBS-LFMP60AH

This battery type is based on single cells with a nominal voltage of 3.2 V and a capacity of 60.0 Ah [LitePower Solutions, 2011]. Therefore, the composition of eight cells in series allows for a nominal voltage of 25.6 V and a capacity of 60.0 Ah.

Summary of Technical Parameters

Relevant technical parameters of the four battery types are summarized in Table 4.4. This overview includes nominal voltages, nominal capacities, weights, and volumes of single cells and the assembled batteries as well as the cell configurations. The 80%-Lifetime parameters define the number of cycles and the amount of energy that can be supplied by a battery until the capacity drops under 80% of the nominal capacity. At this point, the battery is intended to be exchanged by a new one. The cell and battery prizes are approximated values valid in 2011. The last four parameters are calculated from given system parameters. For further information of battery technologies refer to [General Electronics Battery, 2008] or [Soderstrom, 2008].

Evaluation of Battery Technologies

The comparison of the battery technologies is carried out under four criteria: *Adaptability*, *Operation Time*, *Flexibility*, and *Costs*. It is not expected that the criteria *Usability* and *Robustness* are influenced by the type of battery. *Safeness* is also not considered, because it is expected that the cell integration and all required battery monitoring and management functionalities are realized appropriately.

Adaptability (A): The adaptability of a robot system to novel applications is influenced by the battery capacity. Considering the available space for a battery inside a robot, the energy density (Wh/dm³) can be used. Based on Equation 4.8, the weights are 14.8 % (B1), 9.0 % (B2), 54.6 % (B3), and 21.6 % (B4).

Features (F): This criterion follows the same arguments as the *Adaptability* and produces, therefore, the same results.

Table 4.4: *Technical parameters of battery technologies.*

	Alternative	B1	B2	B3	B4
	Chemistry	Lead-Acid	Ni-MH	LiCoO$_2$	LiFePO$_4$
Cell Parameter	Manufacturer	Panasonic	Emmerich	Dynamis	LitePower
	Basis-Cell Type	LC-X1242AP	D9000MAH	LP9675135	LFMP60AH
	Nominal Voltage [V]	12.0	1.2	3.7	3.2
	Nominal Capacity [Ah]	42.0	9.0	10.0	60.0
	Weight [kg]	16.00	0.17	0.22	2.00
	Volume [dm^3]	5.7	0.2	0.1	1.5
	Prize [€]	80	10	20	75
	Configuration (Series x Parallel)	2 x 1	20 x 3	7 x 8	8 x 1
Battery Parameter	Nominal Voltage [V]	24.0	24.0	25.9	25.6
	Nominal Capacity [Ah]	42.0	27.0	80.0	60.0
	Nominal Capacity [Wh]	1,008	648	2,072	1,536
	Weight [kg]	32.0	10.2	12.3	16.0
	Volume [dm^3]	11.4	12.0	6.3	11.8
	80%-Lifetime [Cycles]	300	500	500	1,200
	80%-Lifetime [kWh]	302	324	1,036	1,843
	Prize [€]	160	600	1,120	600
	Energy Density [Wh/kg]	31.5	63.5	168.2	96.0
	Energy Density [Wh/dm^3]	88.6	54.0	326.9	129.7
	Costs-Capacity-Ratio [€/kWh]	158.7	925.9	540.5	390.6
	Costs-Lifetime-Ratio [€/kWh]	0.53	1.85	1.08	0.33

Operation Time (O): Similar to *Adaptability*, the operation time of a robot depends on the battery capacity integrated in a given volume. The maximum charging power of a battery system, limiting the speed of the charging process, is not considered. This is reasonable, because in modern battery systems (with high charging values), the charging speed is usually limited by the capabilities of the power supply units. Therefore, the same weights as calculated for the criterion *Adaptability* can be used.

Costs (C): For the evaluation of the battery costs, the costs-lifetime-ratio is applicable. This value represents the battery costs under consideration of the amount of energy that can be supplied by a battery over its lifetime. This ratio is especially important for applications, in which the lifetime of the robot is higher than the lifetime of the battery. The weights for the described battery technologies are 29.4 % (B1), 8.4 % (B2), 14.4 % (B3), and 47.8 % (B4).

The results of this evaluation process are summarized in Table 4.5.

Table 4.5: Comparison results for battery technologies.

	Technology	A	O	U	R	S	F	C
B1	Lead-Acid	14.8%	14.8%	25.0%	25.0%	25.0%	14.8%	29.4%
B2	Ni-MH	9.0%	9.0%	25.0%	25.0%	25.0%	9.0%	8.4%
B3	$LiCoO_2$	54.6%	54.6%	25.0%	25.0%	25.0%	54.6%	14.4%
B4	$LiFePO_4$	21.6%	21.6%	25.0%	25.0%	25.0%	21.6%	47.8%

Charging System

The intended robot platforms for shopping and home-care applications should be charged in two ways: autonomously and manually. The general operation mode should be the autonomous recharge approach. In this mode, the robot drives to the location of the charging station as soon as the battery is empty. The docking process to the charging station is carried out autonomously without user interference. After the charging process is finished (or in case of an external event, e.g., a user request), the robot docks off from the charging station and continues its normal operations. To compensate for positioning inaccuracies during the docking process of the robot, metal plates with an adequate size can be used for contacting the robot to the charging station. The manual charging mode can be used if no charging station is available or in applications, in which autonomous charging is not required (e.g., trade fairs). To enable the manual charging process, the user has to plug-in the charging connector to the robot and to unplug it to finish the charging process.

For the autonomous charging mode, the transfer of energy from the charging station to the robot can be arranged based on extra-low voltage, line voltage, or electromagnetic induction (Figure 4.6). The transfer of extra-low voltage (C1, C2) provides high safeness for users. The disadvantage of this principle is the higher electrical current that has to be transferred, caused by the lower voltage (assuming an equal performance of all charging technologies). Charging principles based on extra-low voltage are used, e.g., by floor-cleaning robots. In the field of mobile robot research, several development groups apply this concept, like [Silverman et al., 2003] or [Kim et al., 2005].

A charging system based on line voltage (C3, C4) requires significantly less electrical current to transfer the same amount of energy. In this case, it must be assured that persons never get in touch with contacts providing line voltage. The integration of certified power plugs for the transfer of line voltage would solve this problem. Unfortunately, such plugs usually have high demands on the positioning accuracy of a robot. Possible solutions are the integration of additional sensors to better detect the position of the charging station or the integration of mechanical guidance systems to force the robot into the correct position. Another possibility would be the usage of a robot manipulator to execute the docking process [Meeussen et al., 2010].

The third approach uses a contactless transfer of energy based on electromagnetic induction (C5, C6). Such a system allows for higher inaccuracies of the docking process (a range of some centimeters) and is safe for users. Regrettably, inductive charging technologies create higher system costs compared to contact based charging

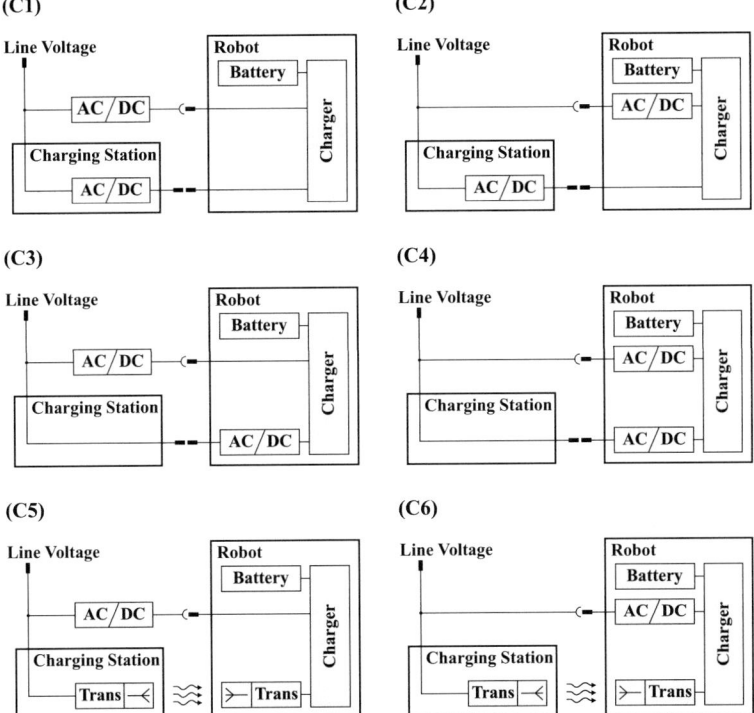

Figure 4.6: *Charging systems for the robot platforms. Every system consists of a manual mode (upper part) and an autonomous mode based on a charging station (lower part). The manual mode provides two versions: an extra-low voltage charging process in combination with an external AC/DC-Converter (C1, C3, C5), and a line voltage charging process in combination with an integrated AC/DC-Converter (C2, C4, C6). The autonomous modes are realized by extra-low voltage (C1, C2), line voltage (C3, C4), or an inductive charging principle (C5, C6).*

principles. An example from the field of robot systems in presented in [Ryan and Coup, 2006].

For the manual charging mode, the integration of an extra-low voltage charging technology or a line voltage charging technology is possible. The difference is the location of the power converter (AC/DC-Converter) either outside the robot (C1, C3, C5) or inside the robot (C2, C4, C6). The decision for the best solution depends on the requirements of the application. A line voltage charging principle allows for a higher usability, because operators do not need external power converters to charge the robot. The realization of an extra-low voltage charging system reduces the costs of the robot system, which can be beneficial for service robots using the autonomous charging mode by default.

Evaluation Process of Charging Systems

Considering the three autonomous and the two manual charging principles, six combinations are possible (Figure 4.6). The following statements and ratings can be derived:

Adaptability (A): The charging system has no impact on this criterion, because all presented solutions are equally applicable to new applications.

Operation Time (O): Service robots are normally charged in the autonomous charging mode, which is exclusively considered for the weighting under this criterion. From experiences it can be assumed that the provided charging power, which influences the operation time, is equal for extra-low voltage solutions (C1, C2) and inductive solutions (C5, C6). The higher amount of transferable power of line voltage based charging principles (C3, C4) essentially increase the charging speed.

Usability (U): The autonomous charging mode does not require any user interference. Still, it should be considered that the usage of an internal power converter (C2, C4, C6) for the manual charging mode is weakly more important for the usability than an external power converter (C1, C3, C5).

Robustness (R): Charging technologies using electrical contacts are assumed to have similar robustness (C1, C2, C3, C4). The main reasons for malfunctions of contact based charging technologies are corrosion, impurity, and deformation. These sources of defects do not apply to contactless charging systems (C5, C6). Therefore, they are weighted as strongly more robust than contact based solutions.

Safeness (S): Every charging technology must prevent the contact of persons with dangerous voltage levels, which is addressed in the manual mode by the application of standard power plugs. For the autonomous charging technologies it should be considered that solutions with extra-low-voltage (C1, C2) are weakly more preferable than solutions based on line voltage (C3, C4). Inductive charging systems provide the highest safeness. Therefore, C5 and C6 are evaluated as weakly more important than C1 and C2, and essentially more important than C3 and C4.

Features (F): The satisfaction of this criterion is not influenced by the applied charging system.

Costs (C): It should be assumed that the costs of available power converters (AC/DC-Converters) are equal to the costs of inductive chargers. Nevertheless, the usage of the autonomous charging technologies C3 and C4 is evaluated as weakly more important than other solutions (C1, C2, C5, C6). The reasons are the lower costs for the charging station, beneficial for applications, in which a robot should be charged at different locations (applicable to big stores or the home environment).

The comparison results for the described evaluation process are presented in Appendix A.2. The calculated weights are summarized in Table 4.6.

Table 4.6: *Comparison results for charging technologies based on extra-low voltage (ELV), line voltage (LV), or inductive transmission (IND).*

	Manual	Auto.	A	O	U	R	S	F	C
C1	ELV	ELV	16.7%	7.1%	8.3%	7.1%	13.0%	16.7%	10.0%
C2	LV	ELV	16.7%	7.1%	25.0%	7.1%	13.0%	16.7%	10.0%
C3	ELV	LV	16.7%	35.7%	8.3%	7.1%	5.3%	16.7%	30.0%
C4	LV	LV	16.7%	35.7%	25.0%	7.1%	5.3%	16.7%	30.0%
C5	ELV	IND	16.7%	7.1%	8.3%	35.7%	31.7%	16.7%	10.0%
C6	LV	IND	16.7%	7.1%	25.0%	35.7%	31.7%	16.7%	10.0%

Drive System

A variety of drive systems are applied in the field of mobile robots. Technical realizations depend on individual requirements of the driving behavior (e.g., speed, maneuverability). Most of the systems can be classified in three groups: systems with overdetermined differential kinematics, systems with differential kinematics and castor wheels, and systems with omni-directional kinematics (overview by [Staab, 2009]). Drive systems with overdetermined differential kinematics were developed for outdoor applications, in which the redundancy of driving wheels are advantageous for the movability on rough terrains. Systems with differential kinematics are often used in indoor applications, where a robust and cost effective solution is needed. Omni-directional kinematics are also used in indoor applications. The advantage of an omni-directional movement is combined with a very high realization effort. The work described in this book focuses on differential platforms with castor wheels, because this technical approach is most applicable for the intended robot applications.

To characterize a differential drive system, three parameters are important. First, the maneuverability of the platform, which depends on the required area of the robot during rotation A_{ROT} compared to the area of the platform footprint A_{PL}. If both values are equal (realized by a circular robot base with a rotation point in the center of the platform), the robot can rotate without the risk of a collision. If the ratio $R_{PL/ROT} = A_{ROT}/A_{PL}$ increases, the turning curve increases and the maneuverability decreases. The second parameter is the stability of the platform. This depends on the stability area created by all wheels A_{ST}. A higher value of A_{ST} means higher stability. The ratio of the stability area to the area of the platform footprint $R_{ST/PL} = A_{ST}/A_{PL}$ describes the saturation of the available platform space for the realization of the stability area. This ratio should be maximal. The center of gravity of the platform should be in the center of gravity of the stability area. The third parameter describes the maximum step height that can be crossed by the platform. This value depends e.g. on the wheels' diameters, the wheels' softness, the power of the motors, or the weight distribution of the platform components.

For the evaluation process, six differential platforms can be considered that are

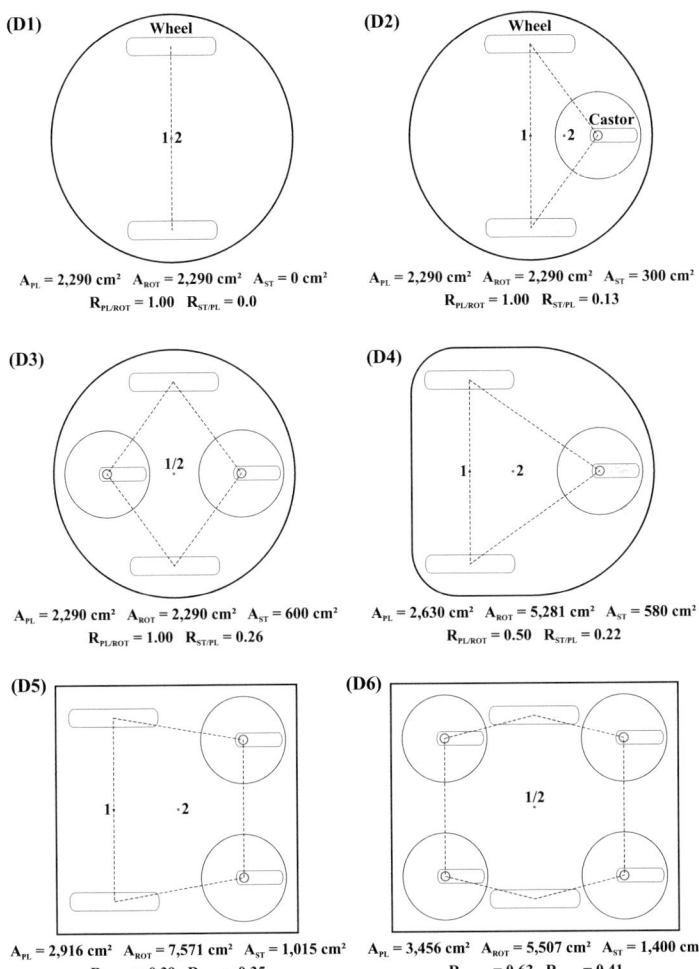

(D1)

Wheel

1｜2

$A_{PL} = 2{,}290 \text{ cm}^2$ $A_{ROT} = 2{,}290 \text{ cm}^2$ $A_{ST} = 0 \text{ cm}^2$
$R_{PL/ROT} = 1.00$ $R_{ST/PL} = 0.0$

(D2)

Wheel

1｜ ·2 Castor

$A_{PL} = 2{,}290 \text{ cm}^2$ $A_{ROT} = 2{,}290 \text{ cm}^2$ $A_{ST} = 300 \text{ cm}^2$
$R_{PL/ROT} = 1.00$ $R_{ST/PL} = 0.13$

(D3)

1/2

$A_{PL} = 2{,}290 \text{ cm}^2$ $A_{ROT} = 2{,}290 \text{ cm}^2$ $A_{ST} = 600 \text{ cm}^2$
$R_{PL/ROT} = 1.00$ $R_{ST/PL} = 0.26$

(D4)

1｜ ·2

$A_{PL} = 2{,}630 \text{ cm}^2$ $A_{ROT} = 5{,}281 \text{ cm}^2$ $A_{ST} = 580 \text{ cm}^2$
$R_{PL/ROT} = 0.50$ $R_{ST/PL} = 0.22$

(D5)

1｜ ·2

$A_{PL} = 2{,}916 \text{ cm}^2$ $A_{ROT} = 7{,}571 \text{ cm}^2$ $A_{ST} = 1{,}015 \text{ cm}^2$
$R_{PL/ROT} = 0.39$ $R_{ST/PL} = 0.35$

(D6)

1/2

$A_{PL} = 3{,}456 \text{ cm}^2$ $A_{ROT} = 5{,}507 \text{ cm}^2$ $A_{ST} = 1{,}400 \text{ cm}^2$
$R_{PL/ROT} = 0.63$ $R_{ST/PL} = 0.41$

Figure 4.7: *Drive systems with differential kinematics. The examples assume a robot width of 540 mm. The characterizing parameters are the area of the platform A_{PL}, the required area of the platform during rotation A_{ROT}, the stability area A_{ST} (dashed lines) described by the positions of the wheels, as well as the ratio $R_{PL/ROT}$ of the platform area to the rotation area and the ratio $R_{ST/PL}$ of the stability area to the platform area. The rotation point of a platform is marked by 1, the optimal point for the center of gravity is marked by 2.*

applicable to the intended service robot applications (Figure 4.7). The drive system D1 consists of just two driven wheels. It has a very good maneuverability and can pass high barriers, because it does not include castor wheels that impede the crossing of barriers. Unfortunately, such platforms are unstable and must be balanced. This worsens the stability of the robot and creates higher costs for motor control and sensor systems. An example is the *Segway* robot platform [Nguyen et al., 2004].

The stability is improved in the drive system D2, which contains additionally a castor wheel at the back side of the robot. Still, this platform can fall over to the front. A (limited) compensation might be a smart placement of system components to bring the center of gravity to the center of the stability area (marked by the number 2 in Figure 4.7). A better solution of this drawback is realized in the drive system D3, which consists of two castor wheels. This platform can still rotate without exceeding the given footprint of the platform. The placement of heavy components (i.e., the battery) is more flexible than in systems D1 and D2. A disadvantage of this platform is the requirement of a spring system. This is necessary for platforms with more than three wheels to always provide a constant surface pressure of the driving wheels. Such a spring system increases the costs for the drive system and decreases the robustness of the platform. An example using the drive system D3 is the robot *Kompai* [Robosoft SA, 2011].

The drive system D4 combines the low-cost realization of a three-wheeled platform with a high stability. Two driven wheels are placed outside the center of the platform's footprint. The disadvantage is that the turning curve appears to be bigger than the footprint of the robot. This has to be considered during the movement of the platform to avoid collisions. An examples for this drive system is the service robot *Charles*, build on a *PeopleBot* platform [Kuo et al., 2008].

The drive system D5 includes two driven wheels at the front side and two castor wheels at the back. Such platforms are usually realized based on rectangular footprints to minimize the required area. The platform D5 allows for a good stability, but has some disadvantages in the maneuverability, because of the significantly enlarged area for rotation A_{ROT}. An example, applying this drive concept, is the rehabilitation robot *FRIEND II* [Volosyak et al., 2005].

The drive system D6 consists of two driven wheels at the center line and four castor wheels placed at the corners of the rectangular footprint. The maneuverability and stability of this platform is better compared to D5, but the complexity and costs for the suspension of the wheels are higher. An example implementation is the *Care-O-Bot I* by Fraunhofer IPA [Graf et al., 2004].

Evaluation of Drive Systems

The following statements and ratings can be derived for the presented robot drive systems:

Adaptability (A): The adaptability of an indoor robot system is not significantly influenced by the drive system.

Operation Time (O): The influence of the drive system on the operation time is based on the power consumption. It is assumed that D1 requires essentially more energy than the other drive concepts, because this platform requires a continuous balancing of the robot. The power consumption of all other platforms should be similar.

Usability (U): The usability depends on the area of the platform footprint A_{PL} and the ratio $R_{PL/ROT}$ during rotation, which should be minimized. Considering these aspects, the drive systems D1, D2, and D3 provide the same usability. These systems are slightly favored over D4, because of the extended area A_{PL} and the worse ratio $R_{PL/ROT}$ of D4. The platforms D1, D2, and D3 are weighted to be "demonstrated" more important than D5 and D6, because of their enlarged platform sizes A_{PL} and the worse ratio $R_{PL/ROT}$.

Robustness (R): The robustness of a drive system is influenced by the complexity of mechanical components that could fail (wheels, bearings, suspensions). The simplest mechanical construction is given by the drive system D1 (two driven wheels). The drive systems D2 and D4 further include a castor wheel that weakly degrades the robustness of a robot platform. The drive systems D3 and D5 consist of a second, spring-mounted castor wheel, which requires addition mechanical components. This strongly decreases the robustness of these platforms in comparison to D1. The worse robustness is given by D6, which requires additional technique for the suspension of several wheels.

Safeness (S): The safeness of a drive system is given by the probability that a robot system tilts over, which depends on the ratio $R_{ST/PL}$ (assuming an optimal distribution of heavy system components). Using Equation 4.9, the safeness of D1 is weighted with 0 %, because this platform is unable to stay without balancing control. Drive system D6 gets the highest safeness weight of 29.9 %, because this version provides the largest stability area compared to the footprint size.

Features (F): This criterion is not influenced by the drive system.

Costs (C): The evaluation of the costs depends on required mechanical and electrical components. It is assumed that version D4 allows for the lowest productions costs, because it does not require a spring system. The systems D1 and D2, which also do not need spring mounted wheels, are weighted to be weakly less important than D4, because both platforms require additional precautions to avoid the tilting over of the robot (e.g., sensors, or a balanced placement of internal components). Because of the integration of spring mounted wheels, the drive systems D3 and D5 are evaluated as essentially, and D6 as "demonstrated" less important than D4.

The results of the pairwise comparison process are summarized in the evaluation matrix (Appendix A.3). Table 4.7 presents the derived weights for the drive systems.

***Table 4.7:** Comparison results for drive systems.*

	Cast.	Shape	A	O	U	R	S	F	C
D1	0	Round	16.7%	3.9%	27.0%	42.0%	0.0%	16.7%	19.9%
D2	1	Round	16.7%	19.2%	27.0%	19.2%	9.5%	16.7%	19.9%
D3	2	Round	16.7%	19.2%	27.0%	8.0%	19.0%	16.7%	8.3%
D4	1	Rounded	16.7%	19.2%	12.1%	19.2%	16.1%	16.7%	38.4%
D5	2	Square	16.7%	19.2%	3.5%	8.0%	25.5%	16.7%	9.2%
D6	4	Square	16.7%	19.2%	3.5%	3.7%	29.9%	16.7%	4.2%

After the weighting of all alternatives, the AHP can now be used to determine the priorities of the alternatives for a dedicated robot system. In the following chapter, the design process of the shopping robot system is described and the outcomes of AHP decisions are presented.

Chapter 5

Development of the Shopping Robot Platform

This chapter describes the development of the shopping robot platform based on the structure of the V-Model. It focuses on the main development activities that were carried out during the second development iteration of this robot platform (Chapter 3.6.4). The first section of this chapter describes the system specification including the overall system specification, the system requirements, and test cases for the evaluation of the final robot system. In addition, this section presents the AHP weighting process for the shopping robot under consideration of the decision criteria (derived in Chapter 4.2.1). Section 5.2 presents the decomposition process of the system into subsystems, segments, and units. It further describes the specifications of subsystems and segments. At this stage, the results of AHP design decisions are presented and discussed in the context of the shopping robot system. Section 5.3 presents the specification of an example system element at unit level and Section 5.4 the realization of this unit. After the design of all system units, these units are composed to build the complete system, which is described in Section 5.5. Finally, testing results of the shopping robot, the compliance of the developed robot system to the system requirements, and applications in several operation areas are presented in Section 5.6.

5.1   System Specification

5.1.1   Overall System Specification

The overall system specification is typically the counterpart to the performance specification of an ordering customer. In this framework, the customer, who is represented by one project partner, defines the needs, whereas the implementing partners evaluate technical solutions to fulfill these requirements. The focus of the overall system specification are the requirements for the SerRoKon shopping robot application.

Initial Situation and Objectives

The structure of stores has changed within the last decades from small downtown shops to huge shopping centers at the brink of the cities. The size of the shopping areas increased with the growing variety of products. Representatives of big retail shops are home improvement stores (Figure 5.1) with up to 60,000 different products distributed over an area from $5,000 \, \text{m}^2$ up to $15,000 \, \text{m}^2$ [Gross et al., 2009]. Home improvement stores often deal with the problem that customers have difficulties to find products in an appropriate time. About 80 % of the customer questions address product locations and prizing information, so stores employees have often to cover the whole sales area [Trabert, 2006]. These questions could be taken over by techni-

Figure 5.1: *Home improvement store.*

cal supporting systems, which would allow the employees to concentrate on more challenging sales conversations. The store would benefit from increased service quality and satisfaction of customers. Further, technical systems would allow for data collection of the customers' behavior, the analysis of buying decisions, and advertisement. Available solutions, e.g., mobile hand held devices, mobile robot solutions, and stationary terminal systems were tested by stores, however, these were not appropriately designed for such challenging tasks and never became accepted by the users.

Functional Requirements

Based on the described situation in shopping centers, the requirements of the new shopping robot platform can be identified. For this purpose, use cases for the application are generated and subtasks determined. These subtasks are used to compose requirements lists to be considered in the development process. In this framework, the focus of these lists is constrained to the robot platform development, even if the use cases also include the description of high level software functionalities.

Use Cases 1 : Customer Interaction

Abstract: The robot shopping guide in a store is supposed to bring customers to requested products and to provide product information about the prize or the availability of the product.

Sequence:

- The robot drives in a given area and attracts customers' attention by speech output.

- The touch screen of the robot is used by interested customers to select required products.

- The robot shows the map of the store and the product location.

- The robot drives to the product location. It avoids collisions with obstacles and persons. On the way, the robot ensures that the user is still following. Changes of direction are indicated by the moveable robot's head.

- After the robot reaches the goal, it offers additional information to the customer. If no further inquiries follow, the robot leaves the customer after a given time and drives back to the starting position.

Subtasks:

- Attention attraction by a pleasant robot design

- Sound replay for customer's attention

- Presentation of information on a touch screen

- Input of information on a touch screen

- Navigation through a given environment

- Obstacle detection and collision avoidance

- Drive deactivation in case of a collision

- Detection of persons around the robot

- Showing of activity, current status, and changes of driving directions by a turnable robot head

- Prize scanning by a barcode reader

- Charging on an autonomous charging station at empty batteries

Use Cases 2 : Operator Interaction

Abstract: The robot shopping guide is stopped by an operator, who checks the current status of the robot.

Sequence:

- An operator of the store authenticates himself at the robot.

- The robot stops its current operation and shows an administration mode.

- The robot can freely be moved by hand.

- Parameters of the robot can be checked (e.g., error status, battery voltage, power consumption).

- Given parameters can be set by the operator (e.g., current location, shut-down time, command for recharging the batteries).

- The operator can reactivate the normal operation mode of the robot.

Subtasks:

- Identification of an authorized operator

- Stopping of all user activities and setting the robot in an idle state

- Analyzing of the robot's current status and parameters

- Setting of system parameters

Requirement Lists

The realization of the described subtasks requires a set of functions that have to be implemented in the robot platform. Every functionality is weighted by a priority (mandatory (M), desirable (D), or optional (O)) that can be considered during the design process. The lists are composed according to functional groups, which simplifies the assignment of subtasks to later development and verification processes. Table 5.1 defines mobility requirements of the platform including the drive system, moving characteristics, and power requirements. Table 5.2 describes functional capabilities and Table 5.3 requirements to the usability of the robot platform. Table 5.4 summarizes requirements for an effective service.

Table 5.1: *Mobility requirements*

Requirement		Prio.	Description
1.1	Stability	M	Stable movement and standing characteristics provided by the mechanical framework and the drive system.
1.2	Motor controller	M	Speed control by the integrated motor controller. Stop of movements in case of a collision or a failure.
1.3	Accurate position	M	Accurate position and movement information of the robot provided by integrated odometry sensors.

1.4	Free run mode	M	A free run mode, in which the robot is easily movable by hand supported by the motor controller.
1.5	Avoiding of collisions	M	Equipment of the robot with a set of sensors to detect obstacles in the environment and to avoid collisions.
1.6	Detection of collisions	M	Detection of collisions by an integrated security sensor.
1.7	Power supply	M	Operation of the robot system with a minimum of eight hours with one battery charge guaranteed by a consumption optimized power supply system.
1.8	Autonomous charging	M	Recharging of the robot by an autonomous charging system without user interference.

Table 5.2: *Functionality requirements*

Requirement		Prio.	Description
2.1	Robot charging	M	Charging by an integrated charging system connected to a standard power plug.
2.2	Charging station	M	Usage of an autonomous charging station to avoid operator interference.
2.3	Pleasant appearance	M	The size and shape of the robot must be pleasant for users.
2.4	Switchable power supplies	O	Software used to switch on and off the power supply of integrated modules.
2.5	Power supply monitoring	D	Monitoring of power outputs to detect faulty modules.
2.6	Barcode reader	O	Usage of an integrated barcode reader for prize scanning.

Table 5.3: *Usability requirements*

Requirement		Prio.	Description
3.1	I/O by touch screen	M	Integration of a touch screen to show information or to modify the robot behavior by touch input.
3.2	Multimedia system	M	Video conferencing and sound output by a set of loudspeakers and microphones.
3.3	Robot head	M	Integration of a robot head to attract users, show basic emotions, and indicate the robot status.
3.4	Head signal lights	M	Usage of a signal light to attract users and to indicate the robot's status.
3.5	Ignition key	M	Authorized power-on by an ignition key.
3.6	RFID reader	O	Integration of a Radio-Frequency IDentification (RFID) reader for access control of an operator.
3.7	WLAN interface	O	Exchange of system information based on wireless communication.

Table 5.4: *Service requirements*

Requirement		Prio.	Description
4.1	Internal errors	M	Detection of internal errors and communication of errors to an operator.
4.2	Mechanical interfaces	O	Realization of a mechanical framework that simplifies the mounting of add-ons.
4.3	Electrical interfaces	O	Easy connection of add-ons by accessible electrical interfaces (communication interfaces, power supplies).

4.4	Accessible internal PC	O	Easy integration of further components or external PC devices to the internal PC by accessible system connectors.
4.5	Transportation	O	Availability of a transportation box to protect the robot during shipping.
4.6	Service PC interface	O	Integration of a service PC interface for system analysis and configuration.

Non-Functional Requirements

This section describes non-functional requirements, relevant for a successful application of the robot. Non-functional requirements include aspects of system quality, performance, service, and maintenance. Moreover, requirements will also be defined for the operation environment of the robot. Where possible, the specification of these requirements includes measurable parameters to validate the implementation quality of these features.

Working Area

The shopping robot platform will be designed for indoor environments. Common environment characteristics are temperature range, humidity, pollution, and floor quality. To account for asperity and bumps, the robot platform should be designed to be able to pass small step heights (15 mm). However, the working area must be free of ledges to protect the robot from falling (alternatively, these areas must be secured).

The working area of the robots in home improvement stores (Figure 5.2) is expected to be very large. Therefore, the robot system should be equipped with a robust drive system, provide an accurate odometry, and achieve appropriate velocities. The usage in stores also requires a reliable detection of obstacles lying on the floor or sticking out of storage racks.

The robot system will interact with walked-in users. Consequently, the operation of the robot has to be user-friendly and intuitive for customers not familiar with robot

TOOM Baumarkt GmbH
Home Improvement Store in Euskirchen

Operational Area: appr. 10.800m²

Figure 5.2: *Map of the Toom BauMarkt home improvement store in Euskirchen.*

technologies. The system design has to consider encounters with elderly or disabled persons. Further, the robot should be robust to manage interactions with children.

Physical Constraints

Physical constrains refer mainly to the size of the robot platform: a small realization would be easily overseen; a large robot might scare people. Project partners agreed that a height of about 1.5 m, corresponding to a twelve years old child would be an appropriate size. For the size of the footprint of the system, there is a trade-off between maneuverability and stability of the platform. A rather small robot base would allow for the passing of narrow passageways, but a wide footprint would increase the stability. The design of the robot would profit from a smart placement of internal components, to bring the center of gravity to a low position to allow for smaller footprints.

The acceptance of a shopping robot is clearly influenced by its appearance. The robot should appear like a smart companion without generating the impact of too much intelligence (to avoid peoples' intentions to start conversations with the robot). Therefore, a cartoon like appearance is preferred for the design of the robot system. A compromise has to be found between a technology driven and a playful driven solution.

87

Reliability and Performance

The operation of a shopping robot requires the availability of the system for a full working day of eight hours. The development of this robot aims to reach non-intermittent working times of more than eight hours by using high power batteries and optimizing the power consumption of the system. The charging time has also to be optimized and is planned to be less than eight hours. The resulting working-to-charging ratio of better than 1.0 would be adequate for most applications.

The system failure rate, caused by technical defects, should be minimized. A down time of less than ten days per year for the first systems, decreasing for the following system generations would be a realistic goal. Security aspects, which also influence the system reliability, are not discussed in this document.

Installation and Service

The robot system has to be developed under the consideration of an easy integration into processes at the customer's side. In particular, for stores, the teaching process during installation, including mapping of the working area or the setup of goal locations should be as simple as possible. The avoidance of additional markers for localization and navigation is planned, so a flexible re-organization of the working area can take place.

A possibility should be provided to check the status of the robot system remotely. In addition, the robot system should be able to check its status autonomously and to communicate upcoming failures (e.g., fading wheels or the aging of the battery) or sudden failures (e.g., blown fuses or the malfunction of a motor) to a service point.

Overall System Architecture and Life Cycle Analysis

This section presents the preliminary design of the overall system architecture and describes first concepts of selected components and interfaces. It further specifies the life cycle of the robot system.

Overall System Architecture

Figure 5.3 shows the concept of the overall system architecture with the main robot components. It consists of embedded control systems (embedded PC, low-level robot control modules), power supplies (battery system, power converters, charging system), sensor systems (vision sensors, distance sensors, emergency-off buttons, collision sensors), human-robot interfaces (touch display, multimedia unit, robot head, signal lights, RFID-reader), and the drive system. A detailed description of the architecture is given in Section 5.2.

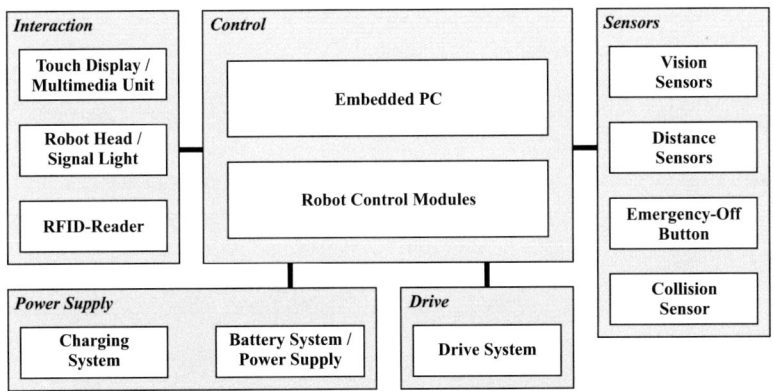

Figure 5.3: *Overall system architecture of the shopping robot platform. This figure represents a functional view of architecture, independent of the later technical realization.*

Interface Overview

The interfaces required by the robot application can be classified into two groups: the human-robot interface, enabling the interaction between users or operators with the robot; and the robot supporting system interface, allowing the communication between the robot and other systems to broaden the robot's functions. The human-robot interaction is mainly realized by the touch screen, microphones, loudspeakers, and the robot head, as well as the RFID reader for administrator access. Users further profit from a barcode scanner to read out prizing information. Operators can use an ignition key to turn on and off the robot. Low level system parameters can be checked and adapted using an optional status display. Most of the functionalities provided by high level software should also be available by remote access to the robot using Wireless Local Area Network (WLAN) to access to the robot's behavior or internal parameters. Thus, service personal can analyze the status of the robot and check for errors online.

A supporting system interface is the control of the autonomous charging station. This interface allows the robot to turn on the charging station, when ready to charge and turn it off as soon as the charging process ends. Further supporting system interfaces depend on the operational environment. Examples are the remote access to an elevator, to autonomously change the floor, or the access to automatic doors.

System Life Cycle

Technological progress, particularly in the fields of personal computing, sensor systems, and battery technologies steadily increases the demand for novel functionalities of the system. To satisfy customers, new development processes have to be initialized after the finalization of the first system. This leads to a life cycle of the robot platform of about three years.

5.1.2 Specification of the Evaluation Process

In order to pass a decision gate, major task outputs of a project stage have to be evaluated and their functionality confirmed. An evaluation specification provides instructions and criteria for the testing procedure of every major task output. To pass the decision gate *System Specified*, the evaluations for two task outputs have to be specified: first, the functional and non-functional robot platform requirements and, second, the content and structure of the overall system specification.

The evaluation specification to verify functional and non-functional requirements provides the system inspector with instructions for testing the functionality of the robot platform and all supporting devices. It contains test scenarios for the verification of all defined system requirements. The evaluation processes are based on black-box-tests, in which evaluation cases simulate expected situations. The failures forced by these tests must be reversible. The test environments are oriented on the later usage. For realistic test scenarios, environments at the manufacturers and the customers side are selected.

The following evaluation case represents a simple example test scenario. This example was chosen, because the charging system is influenced by the AHP decision process. It illustrates that the specification of evaluation procedures at this development stage must focus on required functionaries independent on specific technical realizations (as described in Chapter 4.2.2). A complete description of all evaluation cases is beyond the scope of this document.

Evaluation Case : Charging Procedures

Description: The available robot charging systems are tested.

Environment: Research lab at manufacturer's side.

Initial State: The robot platform is ready for operation. A service PC is connected to the robot to check the status of the internal charging module and the autonomous charging station.

Sequence:

- Connect the internal charging system to line power.

- Set the motor controller to the free run mode by service PC.

- Move the robot platform to its autonomous charging station by hand.

- Activate the internal charging module with the service PC to start charging by the autonomous charging station.

- Interrupt charging by replacing the robot.

Expected Results:

- Robot starts charging after connecting to line power.

- Robot is easily movable after enabling the free run mode of the motor controller.

- Robot can also be charged by the autonomous charging station.

- Robot stops charging by the autonomous charging station as soon as a movement is detected by the motor controller.

Covered Requirements[1]: 1.3, 1.4, 1.8, 2.1, 2.2, 4.6

The evaluation procedures of functional requirements are carried out at the end of the development process (Section 5.6). Non-functional requirements as aspects of quality, performance, service, and maintenance are considered during the development process, however, the verification process of these requirements is part of the decision gate *Acceptance Completed*, which is not in the focus of this work.

The evaluation specification for the overall system specification is based on alternative questions: a *yes* answer indicates the passing of a criterion; a *no* answer requires the revision of this topic. Formal criteria, e.g., questions about document design, orthography, or directory structure are only relevant in case of a separate specification document. The aspects to be covered by this evaluation specification can be found in the V-Model documentation.

5.1.3   Evaluation of AHP Criteria for the Shopping Robot Application

At this stage, the evaluation process of the decision criteria for the interactive shopping robot can be carried out based on the defined functional and non-functional system requirements:

Adaptability (A): The robot system will be primarily developed for shopping and guidance applications. The transfer of this platform in other operation areas (e.g., industry, research) should be easily possible based on the given system characteristics of a shopping robot (e.g., the indoor operation area). Therefore, the criterion *Adaptability* is of small relevance for the development process.

---

[1]Requirements as defined in the requirement lists (Section 5.1.1).

Operation Time (O): The operation time of the robot with one battery charge should be at least eight hours. A continuous availability over the whole opening period of a store is not mandatory. Based on prior estimations of the battery capacity and the power consumption of (adequate) system components, it is assumed that this requirement can be fulfilled by a reliable development process. Special efforts during the system design process seem not to be required. Consequently, the satisfaction of this criterion is of low priority. It is weighted to be weakly less important than *Adaptability*.

Usability (U): The usability of the robot is necessary for the success of the shopping robot application. It must be considered that customers in stores are not familiar with the usage of robot systems. Additionally, a good operability by the store's employees should also be taken into account. The *Usability* is weighted to be essentially more important than *Adaptability* and *Operation Time*.

Robustness (R): The asperity of the operation area (e.g., small bumps, tile joints) and the interaction with persons (i.e., children) require a high robustness of the system. A low system failure rate is aspired. This criterion is evaluated as equal important as *Usability* and essentially more important than *Adaptability* and *Usability*.

Safeness (S): The robot system must guarantee absolute safety for people. The development process has to be carried out under the consideration of an adequate safety concept. Preferably, sensor systems complying with safety standards should be integrated. These precautions are especially important, because of the usage in public areas. This criterion was weighted to be essentially more important than *Adaptability*, *Operation Time*, and *Usability*. It is further weakly more important than *Robustness*.

Features (F): This criterion is considered, if a robot is planned to be used for further developments and applications. For the realization of the shopping robot application, integrated features that go beyond the system requirements are not of relevance. This criterion is weighted to be equal important as *Adaptability*.

Costs (C): The robot should be designed under aspects of production and service costs. This is essential for marketing and a wide distribution of this system. Acceptable costs for a store depend on the benefit of the robot system. This criterion is weighted to be essentially more important than *Adaptability*, *Operation Time*, and *Features*; and essentially less important than *Usability* and *Safeness*. It is equally important as *Robustness*.

The results of the pairwise comparisons and the calculated weights are presented in Tables 5.5.

Table 5.5: *Criteria evaluation matrix of the shopping robot platform. It shows the pairwise comparison results of the criteria and the calculated weights based on the equations of Chapter 4. The Consistency Ration (C.R.) shows the sufficient consistency of this matrix.*

	A	O	U	R	S	F	C	Weights	C.R.
A	1/1	3/1	1/3	1/3	1/5	1/1	1/3	**6.7 %**	
O	1/3	1/1	1/3	1/3	1/5	1/3	1/3	**4.2 %**	
U	3/1	3/1	1/1	1/1	1/5	3/1	3/1	**17.0 %**	
R	3/1	3/1	1/1	1/1	1/3	3/1	1/1	**14.6 %**	0.061
S	5/1	5/1	5/1	3/1	1/1	5/1	3/1	**37.4 %**	
F	1/1	3/1	1/3	1/3	1/5	1/1	1/3	**6.7 %**	
C	3/1	3/1	1/3	1/1	1/3	3/1	1/1	**13.4 %**	

For the illustration and comparison of evaluation process outcomes, chart diagrams can be used. Figure 5.4 presents the criteria weights for the shopping robot platform.

5.2   System Design

This section describes the robot design process at the hierarchical levels of systems, subsystems, and segments. (Figure 5.5). The identified system architecture is presented and the system decomposition process is carried out.

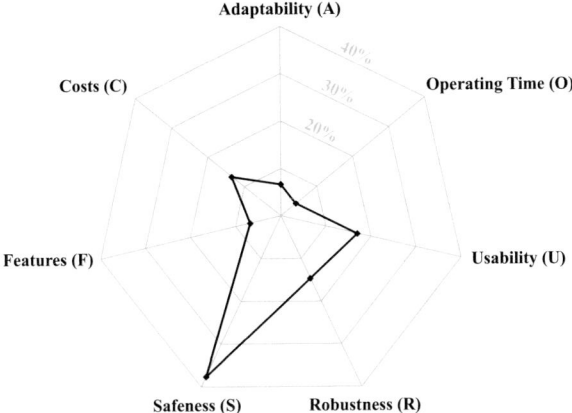

Figure 5.4: *Criteria chart for the shopping robot platform representing the calculated criteria weights of Table 5.5.*

To pass the decision gate *System Designed*, subsystems and segments have to be specified. This chapter defines the characteristics of identified subsystems and segments and highlights the AHP design decisions. The specification of segments containing the system overview, the interface specification, non-functional requirements, and an evaluation specification is not included in this chapter, because of the given system complexity at these levels and the focus of this work on the principle development course. A detailed description of the requirements of the subsystem and segment specifications can be found in the V-Model documentation.

5.2.1 System Architecture

An appropriate system architecture has to be chosen for the robot platform to allow for the realization of all functional and non-functional requirements. The weighted system architectures, presented in Chapter 4.2.2, are ranked according to the weighted criteria for the shopping robot. The resulting priorities for the described system architectures are presented in Table 5.6.

Table 5.6: *Decision results for the system architectures.*

	A 6.7%	**O** 4.2%	**U** 17.0%	**R** 14.6%	**S** 37.4%	**F** 6.7%	**C** 13.4%	**Prio.**	**Rank**
A1	12.5%	36.4%	25.0%	5.4%	5.7%	8.3%	43.4%	15.9 %	**4**
A2	12.5%	36.4%	25.0%	14.6%	26.3%	8.3%	43.4%	25.0 %	**2**
A3	12.5%	6.6%	25.0%	23.7%	12.2%	41.7%	4.0%	16.7 %	**3**
A4	62.5%	20.7%	25.0%	56.3%	55.8%	41.7%	9.2%	42.4 %	**1**

The AHP prioritization results reveal that the system architecture A4 (Figure 4.5) fulfills the weighted criteria best. This architecture, composed by a main control unit (embedded PC) and several decomposed control modules, provides a high robustness and safeness, because of the redundancy of system functionalities [Merten and Gross, 2008]. These criteria were crucial factors for the choice of this architecture. The higher system costs and the reduced operation time, caused by the system modularity, had minor influences on the decision results.

5.2.2 System Decomposition

The system decomposition is a hierarchical process, which successively breaks down the system into lower abstraction levels. This process starts with the segmentation of the system into subsystems that constitute the main functional elements of the whole system. The following step is the decomposition of the subsystems into segments, which represent functional groups. Segments usually compose modules with similar functionalities or physical locations in the system. The final step is the breakdown of segments into units. Every unit covers a set of functionalities and can be classified as software units, hardware units, embedded system units, or external units, whereas external units are the purchased parts of the system.

In the following, the decomposition process of the shopping robot is described, focusing on subsystem, segment, and unit levels (Figure 5.5). A further decomposition of units into components is described in Section 5.3.

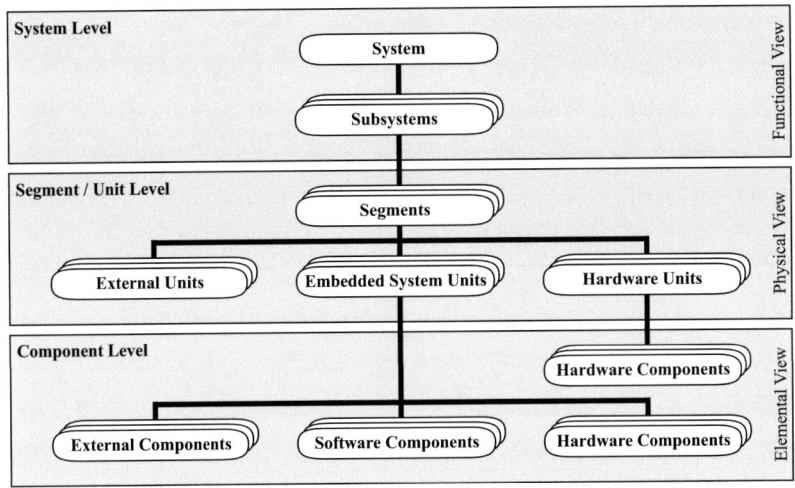

Figure 5.5: *System architecture hierarchy of the robot system.*

Decomposition of the System into Subsystems

Figure 5.6 shows the decomposed system of the robot platform. The mobile robot system requires the *Control Subsystem* for the control and monitoring of the robot's functionalities, the *Power Supply Subsystem* for the provision of the system energy and the recharging of the energy storage, the *Drive Subsystem* for the movement of the robot, the *Sensor Subsystem* for the analysis of the robot's environment, and the *Interaction Subsystem* for the communication with users and operators.

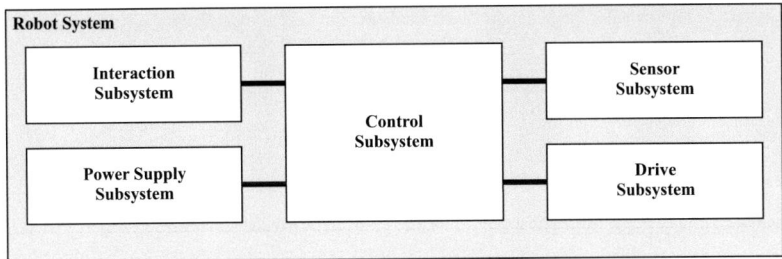

Figure 5.6: *Decomposition of the robot system into subsystems.*

Decomposition of Subsystems to Segments

The second step is the breakdown of the identified subsystems into segments. These segments associated with the related subsystems for the robot platform are depicted in Figure 5.7.

Control Subsystem

High level algorithms for navigation, localization, or human-machine interaction are computed on the *Main Control Segment*, which is realized by an embedded PC. The advantages are the high computational power, the storage of large data sets, and a flexible adaptation of software functionalities. However, this system needs a high amount of energy to process all information. Thus, a power consumption optimized PC is integrated.

To account for an optimal realization of real-time tasks and determined reaction times, the *Control Subsystem* includes a *Control Module Segment* that is optimized for real-time low-level control. This segment consists of small modules for the control of other segments. For example, a motor controller for the stimulation of the *Drive*

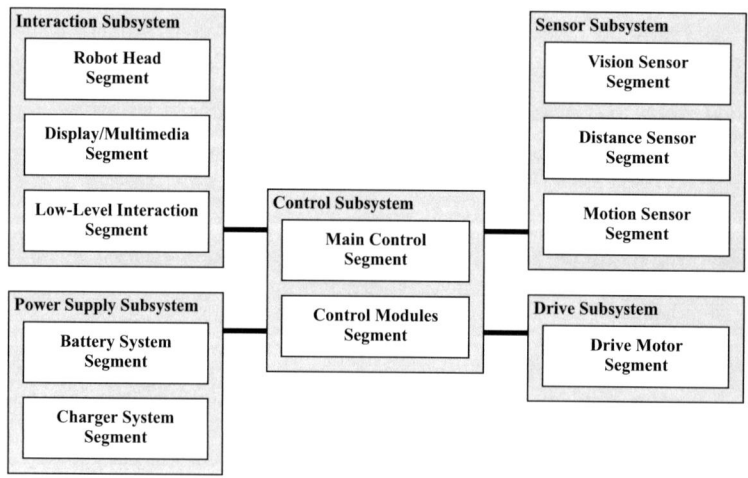

Figure 5.7: *Decomposition of subsystems to segments*

Motor Segment, or a sensor module for the data exchange with the *Distance Sensor Segment* or the *Motion Sensor Segement*.

Power Supply Subsystem

This subsystem contains the *Battery System Segment*, including the robot's battery and a module, which is mainly responsible for battery monitoring; as well as the *Charger System Segment*, including technical systems for the battery charging process. The realizations of both segments are affected by design decisions that were carried out based on the AHP.

At the beginning of the shopping robot development process, the only battery technologies, which came into question, were lead-acid and Ni-MH batteries. The lead-acid battery with a nominal voltage of 24.0 V and a nominal capacity of 42.0 Ah was integrated, because of the higher energy density and the lower costs (see Table 4.4).

Table 5.7: *Decision results for the battery systems.*

	A	O	U	R	S	F	C	Prio.	Rank
	6.7%	4.2%	17.0%	14.6%	37.4%	6.7%	13.4%		
B1	14.8%	14.8%	25.0%	25.0%	25.0%	14.8%	29.4%	**23.8 %**	3
B2	9.0%	9.0%	25.0%	25.0%	25.0%	9.0%	8.4%	**19.9 %**	4
B3	54.6%	54.6%	25.0%	25.0%	25.0%	54.6%	14.4%	**28.8 %**	1
B4	21.6%	21.6%	25.0%	25.0%	25.0%	21.6%	47.8%	**27.5 %**	2

In an additional development cycle (after the finishing of the SerRoKon projects), novel lithium battery technologies were available. At this stage, the AHP was used to evaluate the now four possible alternatives (described in Chapter 4.2.2). Table 5.7 presents the results of this decision process. A lithium-polymer battery based on $LiCoO_2$ cells (alternative B3) was selected as most suitable for the shopping robot application. The battery configuration, used for the evaluation process, consisted of 56 cells (seven in series, eight in parallel) and provided a nominal voltage of 25.9 V and a nominal capacity of 80.0 Ah. The advantage of this battery type is the very high energy density in comparison with the other battery types. The main disadvantage is the prize.

Similar to lithium batteries, inductive charging technologies were not available during the development process of the shopping robot in the SerRoKon projects. Initially, the AHP process was carried out with the alternatives C1 to C4 leading to the ranking: C2 (29.4 %), C1 (25.2 %), C4 (24.8 %), and C3 (20.6 %). Therefore, the charging system C2 was integrated into the robot. This system consists of an autonomous charging station with integrated power converter, which provides extra-low voltage for the charging of the robot, and a manual charging mode based on line voltage. The realization of the charging system C2 is presented in the project stage *System Integration* (Section 5.5).

After inductive charging technologies were available, the AHP process was repeated based on all alternatives. Table 5.8 shows the results of this process and reveals that the charging system C6 (an inductive charging system in combination with a line voltage manual charging mode) is most applicable to the shopping robot application. This charging system will be integrated into the shopping robot platform in 2012.

Table 5.8: Decision results for the charging systems.

	A 6.7%	O 4.2%	U 17.0%	R 14.6%	S 37.4%	F 6.7%	C 13.4%	Prio.	Rank
C1	16.7%	7.1%	8.3%	7.1%	13.0%	16.7%	10.0%	**11.2 %**	6
C2	16.7%	7.1%	25.0%	7.1%	13.0%	16.7%	10.0%	**14.0 %**	4
C3	16.7%	35.7%	8.3%	7.1%	5.3%	16.7%	30.0%	**12.2 %**	5
C4	16.7%	35.7%	25.0%	7.1%	5.3%	16.7%	30.0%	**15.0 %**	3
C5	16.7%	7.1%	8.3%	35.7%	31.7%	16.7%	10.0%	**22.3 %**	2
C6	16.7%	7.1%	25.0%	35.7%	31.7%	16.7%	10.0%	**25.2 %**	1

The results of Table 5.8 disclose a disadvantage of the AHP: The ranking of alternatives might change, when alternatives are added to or removed from the AHP decision process. This effect is further discussed in Chapter 7.

Drive Subsystem

The *Drive Motor Segment*, the only element in this subsystem, determines the drive system (wheel configuration, platform footprint) of the robot. The decision on an adequate drive concept is based on the AHP (results summarized in Table 5.9).

Table 5.9: *Decision results for the drive systems.*

	A 6.7%	O 4.2%	U 17.0%	R 14.6%	S 37.4%	F 6.7%	C 13.4%	Prio.	Rank
D1	16.7%	3.9%	27.0%	42.0%	0.0%	16.7%	19.9%	**15.8 %**	5
D2	16.7%	19.2%	27.0%	19.2%	9.5%	16.7%	19.9%	**16.6 %**	3
D3	16.7%	19.2%	27.0%	8.0%	19.0%	16.7%	8.3%	**17.0 %**	2
D4	16.7%	19.2%	12.1%	19.2%	16.1%	16.7%	38.4%	**19.1 %**	1
D5	16.7%	19.2%	3.5%	8.0%	25.5%	16.7%	9.2%	**15.6 %**	6
D6	16.7%	19.2%	3.5%	3.7%	29.9%	16.7%	4.2%	**15.9 %**	4

Given these priorities, the robot platform is designed based on the drive system D4. The decision for this version was primary caused by the adequate costs and a good robustness of this concept. The selected drive system consists of two driven wheels and one castor wheel at the backside. The roundish shape is also applicable to the shopping robot application.

Interaction Subsystem

This subsystem is composed of the *Display/Multimedia Segment*, the *Robot Head Segment*, and the *Low-Level Interaction Segment*. The *Display/Multimedia Segment* is the communication interface between the robot and the user. It mainly consists of the touch screen, loudspeakers, and microphones. The touch screen can be used to present information and to receive user inputs. The robot can generate sound or speech outputs by its loudspeakers. The integrated microphones can be used for video conferencing.

A robot head is installed on the top of the robot platform to create a cartoon-like appearance and to show movement intentions and basic emotions to the user. The head and the integrated eyes can be rotated or tiled. The eye lids can be opened and closed to generate the impression of winking or sleeping. At the top of the head, a signal light is integrated that shows the current robot status.

The *Low-Level Interaction Segment* contains additional input and outputs devices, primarily for the interaction with operators. One examples is the RFID reader that can be used to set the robot application in an administrator mode.

Sensor Subsystem

This subsystem includes three segments for different types of sensors: the *Distance Sensor Segment*, the *Motion Sensor Segment*, and the *Vision Sensor Segment*. The *Distance Sensor Segment* consists of a laser range finder and ultrasonic sensors. The laser range finder is integrated to detect obstacles and persons for collision avoidance. It is further used by localization algorithms to compute the current position. The time of flight of the sensor signal reflected by an object is the information for the calculation of the object distance. The physical constraints due to mirrored or transparent surfaces are addressed by the integration of ultrasonic sensors as a redundant system. These sensors produce an acoustic beam that is better reflected by glass surfaces. Ultrasonic sensors also provide a higher beam angle, which is necessary to detect flat objects on the floor.

The *Motion Sensor Segment* includes sensors that are required for a reliable and safe motion of the robot platform. This segment contains a bumper that detects collisions. This sensor works like a button that is pressed as soon as the robot hits an object. In this case, the motor controller has to stop the drive system immediately. A similar functionality is given by optional emergency buttons that can be activated by a user.

Vision sensors of the *Vision Sensor Segment* are integrated for multiple reasons. First, these sensors can be used to detect persons in the robot's environment, which is important to realize a smart robot behavior. Second, they can be used for the detection of high obstacles that protrude in the way of the robot. Vision sensors can further be used during video conferencing to show a picture of the robot's surrounding.

Decomposition of Segments into Units

The breakdown of segments into units is the final step of the system decomposition process in this section. In addition to the three standard units suggested by the V-Model (hardware, software, and external units), a fourth standard unit for embedded systems is defined to simplify the decomposition process (the capital letters in brackets are used in the figures):

Hardware Unit (H): This group includes hardware elements composed of hardware components without software. Examples are connector boards, toggle switches, or buttons.

Software Unit (S): This group combines exclusively software elements. Software units are composed hierarchically of software components.

External Unit (X): External units are not developed in the scope of this project. They include off-the-shelf products or modules developed in advance. An *External Unit* may include hardware and software components.

Embedded System Unit (E): This group consists of modules that integrate software and hardware components, which might be further decomposed. Examples are uC based control modules, embedded PCs, or sensor systems.

The assignment of system units to segments is ambiguous for some components. For example, the barcode reader belongs to the *Low-Level Interaction Segment*, because a user receives product information after scanning the product's bar code. It can also be classified as an element of the *Vision Sensor Segment*, because of the camera based sensor concept. The final decision for the best fitting segment of a unit depends on the system designer.

In the following, the decomposition of segments into units will be described. The amount of different segments makes it impossible to discuss all units in this document. Therefore, three example segmentations that are most relevant for the system development process will be presented. Further segmentations can be found in Appendix B.

Main Control Segment

The *Main Control Segment* (Figure 5.8) contains the *Embedded PC Unit*, a *USB-CAN-Converter Unit*, and the *Interface Unit*. The first unit is equipped with an embedded PC to run high-performance software algorithms. This embedded PC is connected to other units by standard communication interfaces, e.g., a Low Voltage Differential Signaling (LVDS) interface to the display, an RS232 interface to the touch sensor of the display, or USB interfaces to the multimedia and camera system.

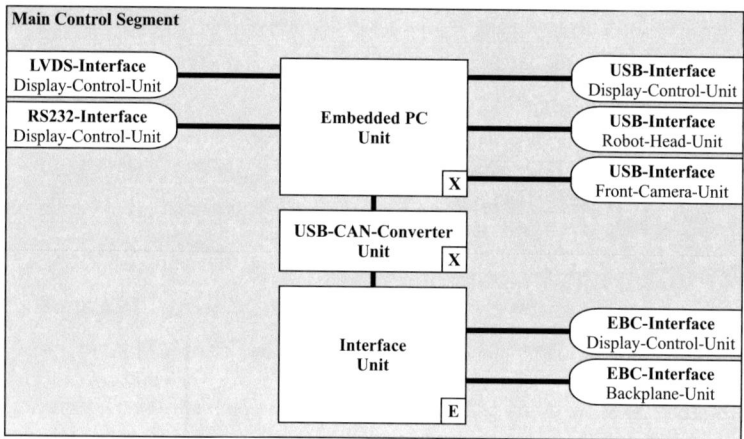

Figure 5.8: *System decomposition of the Main Control Segment.*

The *USB-CAN-Converter Unit* represents the gateway between the *Embedded PC Unit*, connected by USB, and the *Interface Unit*, providing the data exchange to other control units of the robot by the CAN bus.

Control Modules Segment

The second segment of the *Control Subsystem* is the *Control Modules Segment* (Figure 5.9). It is planned to contain several control units: The *Power Control Unit* is designated to connect the incoming power and communication channels of the *Battery Control Unit* with the *Backplane Unit*. The *Backplane Unit* is a hardware interface, where other control units can be plugged in to be connected to the robot's power supply and communication interfaces. The *Sensor Control Unit* is responsible for the control of the ultrasonic sensors. It generates the power supply for these modules and controls the exchange of the sensor information.

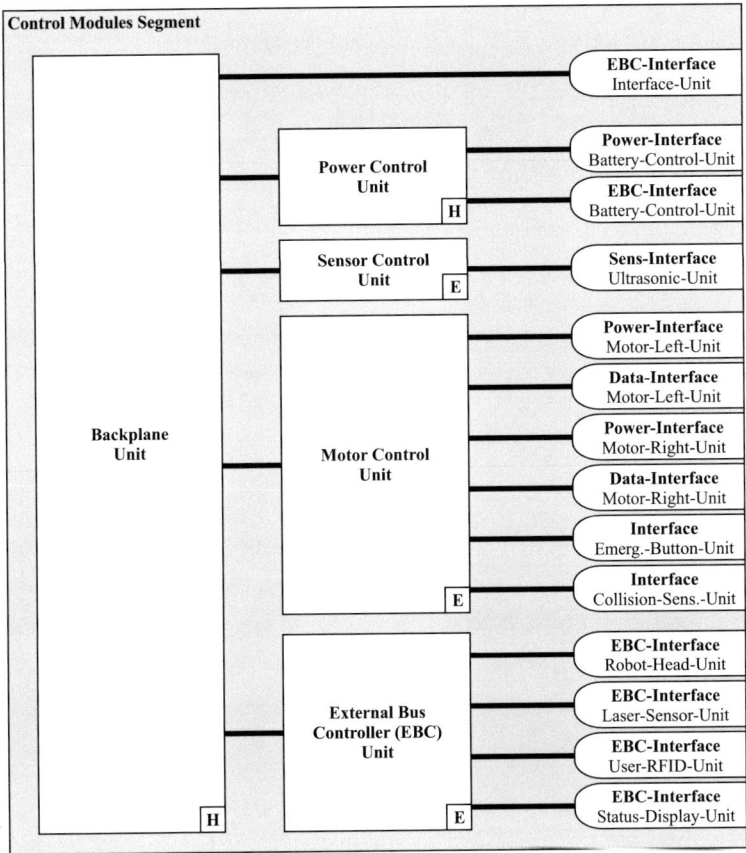

Figure 5.9: *System decomposition of the Control Modules Segment.*

The *Motor Control Unit* controls the movement of the platform in real-time, generates necessary power signals for the motors, and analyzes incoming incremental sensor information. This unit computes the robot's odometry. Security sensors, e.g., the collision sensor or optional emergency buttons are directly connected to this module to stop the movement of the motors in case of a collision or user interference.

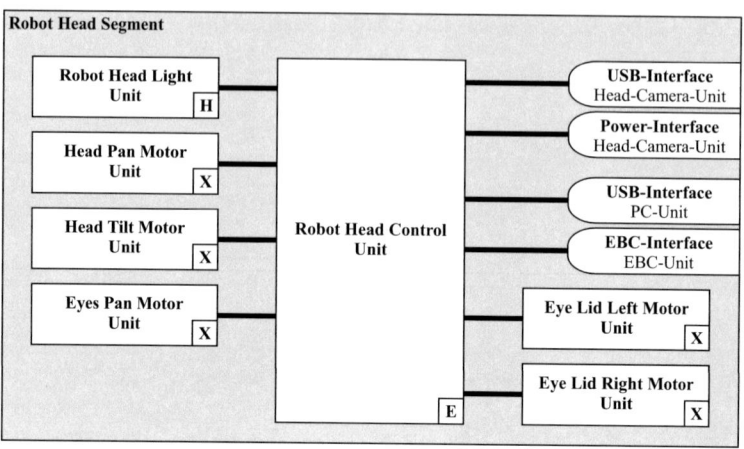

Figure 5.10: *System decomposition of the Robot Head Segment.*

The *External Bus Connector (EBC) Unit* generates power voltages for connected modules like the *Robot Head Unit*. The voltage levels are 5.0 V, 12.0 V, and about 25.9 V (battery voltage). It should be possible to turn on and off every voltage channel separately and to monitor for over-current. This unit provides the EBC-interface for connected devices containing all generated power levels and a CAN bus interface.

Robot Head Segment

The *Robot Head Segment* (Figure 5.10) includes the *Robot Head Control Unit*, which controls the movement of all motor units of the robot head, the *Robot Head Light Unit*, and the *Head Camera Unit*. The *Robot Head Control Unit* consists of a uC, which is running the firmware, and an FPGA, which is generating fast output signals (e.g., PWM outputs).

A *Robot Head Light Unit* is integrated to show the status of the robot by an LED array. Several *Motor Units* are integrated to move the mechanical parts of the robot head. Possible movements are: rotating and tilting of the head, rotating of the eyes, and opening and closing of the eye lid independently.

5.2.3 Interface Overview

Communication interfaces allow for the data transition between system elements. Depending on the information to be exchanged, specific requirements exist for each communication channel. For example, video or sound signals require high-speed transfers; information about the battery status or a scanned barcode can be transmitted at lower speeds. The following interfaces are implemented for the interconnection of system units:

Controller Area Network (CAN) Interface

The CAN interface was developed by the company Bosch for the usage in the automotive sector [Robert Bosch GmbH, 1991]. It belongs to the group of field buses and supports an asynchronous, serial data exchange with data rates up to $1\,\text{MBit/s}$. A high transmission reliability is realized by the integration of redundancy based on bit stuffing and a CRC checksum. CAN bus nodes can be accessed based on assigned message identifiers. The CAN protocol defines an addressing range of 2^{11} identifiers in the standard format and 2^{29} identifiers for the extended format. The CAN interface uses the Carrier Sense Multiple Access/Collision Avoidance (CSMA/CA) method is used to detect data collisions, if multiple communication nodes try to send messages at the same time. This communication principle allows that messages with lower identifiers have a higher probability for a successful transmission.

The CAN bus is the general communication interface for the data transfer between low-level system units, like the *Motor Control Unit*, the *Display Control Unit*, or the *Head Control Unit*. It is connected to the embedded PC by a USB-CAN-Converter. The signals of the CAN bus are combined with switchable power outputs to the EBC-interface to simplify the integration of additional modules.

Inter-Integrated Circuit (I2C) interface

The I2C interface, developed by the company Philips Semiconductors, is used for the inter-communication between circuits on a Printed Circuit Board (PCB) [Philips Semiconductors, 2000]. It is a synchronous communication system based in the Serial Data Line (SDA) and Serial Clock Line (SCL) signals. The communication requires a master to initialize the communication process and to generate the clock signal. Different slaves can be accessed based on a seven bit address space. To allow for the connection of multiple slaves, open-collector outputs are integrated in combination with pull-up resistors. This requires low speed communication frequencies of less than 400 kHz.

The I2C bus is applied for the communication between the *Sensor Module* and the *Ultrasonic Sensor Modules*. This is reasonable, because the low data rates of these sensors do not require the integration of the more complex CAN bus. This enables the usage of small uCs, which reduce the system complexity, the power consumption, and system costs.

RS232 Interface

This interface is used for low speed and low cost communication devices. It was introduced by the Electronic Industries Alliance (EIA) in 1960. The RS232 communication interface is an asynchronous serial data interface and consists of two unidirectional data signals and multiple control signals (e.g., for hardware handshake). The communication is byte oriented and can be extended with a parity bit to improve the transmission reliability. The communication speed can reach 500 kBit/s.

The RS232 communication interface is available for devices that still support this communication interface. For example, RS232 is used to connect the touch sensor of the display to the embedded PC.

Universal Serial Bus (USB) Interface

The USB interface was developed for the connection of PCs to peripheral devices [Compaq et al., 2000]. It consists of two twisted pairs of signals and power supply lines. Every USB port can be connected to one device. An extension of available USB ports can be achieved by USB hubs that distribute one input port to multiple output ports. The resulting tree structure of this communication system allows up to 127 devices to be connected to a PC. Possible transition speeds are 1.5 MBit/s for low speed, 12 MBit/s for full speed, and 480 MBit/s for high speed devices. The power supply of the USB port provides 5.0 V with up to 500 mA.

For the robot platform, this interface is used for the connection of signal converters and standard PC devices to the embedded PC. An example is the USB-CAN-Converter for the access of the embedded PC to the CAN bus.

Low Voltage Differential Signaling (LVDS) Interface

The LVDS interface was developed for the transfer of high speed information. It is based on differential signals with a typical signal swing of 350 mV [National Semiconductor, 2008]. Parallel data signals, synchronized to a clock signal, allow for maximum data rates of up to 3 GBit/s.

LVDS is used for the interface of the embedded PC to the touch display of the robot. The integration of this communication provides a flexible choice of available displays, because most displays support the LVDS interface.

5.3 Detailed System Design

In this development stage, the hardware and software architectures of the system units have to be created. Functional and non-functional requirements of the units have to be derived from higher hierarchical levels of the system and an evaluation specification for the later testing of the realized units must be prepared. In this document, the development process at unit level is exemplarily described on the example of the design of the *Robot Head Control Unit*. The high complexity and the various interfaces of this unit make it suitable for the detailed description of the system design process. The realization of this unit is presented in Section 5.4.

Based on the system decomposition process, the *Robot Head Control Unit* is responsible for the control of all components of the robot head (Figure 5.10). This includes several stepper and Direct Current (DC) motors for the movement of head components (e.g., the eye lids) and the *Robot Head Light Unit*. An optional omnidirectional camera, placed on the top of the robot, should also be connected to this unit, which requires data interfaces and a power supply. The *Robot Head Control*

Unit has to handle the following functionalities:

- Control of four stepper motors.

- Control of one DC motor.

- Output of eight PWM signals for the LED-matrix of the *Robot Head Light Unit*.

- Power supply and data interface (USB) for the omni-directional camera.

- Data interface to other control units (CAN).

5.3.1   Hardware Architecture and Specification

Given this variety of functionalities, a hardware architecture has to be chosen that allows to generate synchronized output signals (motor control, LED control), to process incoming information (motor positions), to communicate with other control units (CAN bus), and to handle analog signals (power supply of the omni-directional camera). There are three possible solutions for the control of these functionalities: the integration of a powerful uC that is able to process fast signals; the integration of an FPGA with an Intellectual Property Core (IP-Core) for the implementation of software algorithms; or the combination of a low-power uC for software algorithms with a small FPGA for the processing of fast signals.

The hardware architecture of the *Robot Head Control Unit* is based on a combination of a uC and an FPGA (Figure 5.11). The uC executes software algorithms, communicates with other control units, and monitors the analog signals. The FPGA generates all signals for the control of the DC motor, one stepper motor, and the LEDs. For the control of the other three stepper motors, special motor control Integrated Circuits (ICs) are used that can generate the feeding currents of the motors in real-time.

In comparison to an architecture with a single powerful uC, this architecture allows for the parallel processing of input and output information using an additional

110

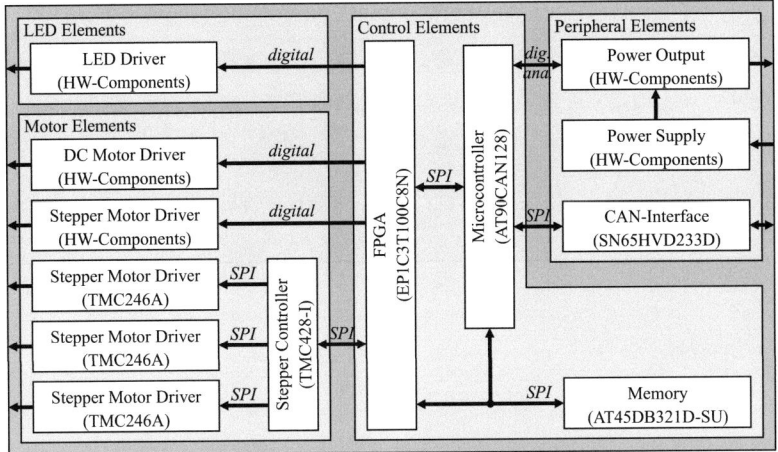

Figure 5.11: *Hardware architecture of the Robot Head Control Unit.*

FPGA. Compared to a single FPGA architecture, this solution provides a higher adaptation flexibility of software functionalities (bootloader capability) and requires less power than an FPGA with integrated IP-Core. Moreover, the usage of a uC, that is also applied in other control units, allows for the re-usage of developed software algorithms.

Hardware Element Overview

Control Elements

An 8-bit uC is integrated as the main control element of the *Robot Head Control Unit.* It executes all software algorithms and controls the functionalities of all unit elements (by means of the FPGA). The applied type of uC must allow for the implementation of all required software and communication functionalities, like the handling of the CANopen communication stack for data exchange with other control units (e.g., the embedded PC) or the control and monitoring of the power output of the omni-directional camera.

The FPGA is used for the control of fast-changing signals. This includes the generation of digital output signals for the DC motor and one stepper motor. For the control of the other stepper motors, the FPGA provides an Serial Peripheral Interface (SPI) gateway between the uC and the *Stepper Controller*. Another functionality of the FPGA is the generation of eight PWM signals for the brightness control of the LED matrix.

The *Memory* stores the configuration data of the FPGA. This memory can be programmed by the uC, which enables the programmability of the FPGA inside the system.

Peripheral Elements

Peripheral Elements include components for the CAN communication, the power supply, and the power output for the omni-directional camera. The components of the *CAN Interface* include a driver IC that converts the Transistor-Transistor-Logik (TTL) signal levels of the uC to differential signals, used by the CAN bus. The *Power Supply* converts the incoming supply voltage to required voltage levels of all unit components. The *Power Output* contains components for the powering and monitoring (voltage and current) of the omni-directional camera.

Motor Elements

The five motors of the robot head are connected to the *Motor Elements*. These elements contain electrical components for the generation of required voltage and current sequences. The DC motor (head rotation) and one stepper motor (head tilting) are directly controlled by the FPGA, whereas the other stepper motors (eye tilting and eye lids) are connected to *Stepper Motor Drivers*. The reason for the integration of special motor drivers is the high acceleration required for these motors (to initiate natural movements). The real-time control sequences of these motor drivers are generated by the *Stepper Controller*.

LED Elements

The *LED Driver* consists of Metal-Oxide-Semiconductor Field-Effect Transistors (MOSFETs) that amplify the output signals of the FPGA to drive the LEDs of the *Robot Head Light Unit*.

Interface Specification

Interfaces can be classified into external interfaces between units and internal interfaces between system components. The *Robot Head Control Unit* supports several types of external interfaces for the interaction with other units. The connection of the *Robot Head Control Unit* to other control units of the robot is based on the EBC-interface (containing the CAN bus and voltage levels of 5.0 V, 12.0 V, and about 25.9 V). An incoming USB interface is directly connected to the omni-directional camera port. This camera port further contains the switchable power output of the camera supply voltage. For the connection of all motors and the LED-matrix, several output lines are used to supply these components with the required supply currents. The implementation of these outputs (configuration, voltage levels, current levels, connector types) has to be realized according to the requirements of the external units.

For the internal exchange of information between system components (uC, FPGA, *Memory*, and the *Stepper Controller*), the SPI interface is applied. This interface has been developed by Motorola to provide a simple interface between ICs. It uses four data lines (chip select, clock line, and two uni-directional data lines) for a synchronized data transfer [Chattopadhyay, 2010]. Furthermore, several digital and analog interfaces must be integrated for the control of system elements.

A detailed description of the data and signals, required for the V-Model development process, is not in the focus of this document.

Non-Functional Requirements

Technical Constraints

Th design of the *Robot Head Control Unit* has to be carried out under the consideration of system requirements given by higher hierarchical levels. For example, the applied battery technologies require that the *Robot Head Control Unit* is able to operate with an input voltage range from 22.0 V to 30.0 V. The working area of the shopping robot is of relevance for the component selection process, because it defines the required ranges for temperature and humidity. The usage in public areas requires the compliance of all electric systems with the technical standards for Electromagnetic Compatibility (EMC) and Electromagnetic Interference (EMI).

The selection process of unit components should also follow two requirements: the re-usage of applied components and the consideration of costs and power consumption. The re-usage of components allows for a higher efficiency of the development process, because designed hardware elements and written software code can be applied to other units. Costs and power consumption of system elements should be considered during the selection process, even if the criteria *Costs* and *Operation Time* are not of highest priority for the shopping robot system.

Usability and Maintainability

The realization of this unit should consider aspects of maintainability. This includes the possibility to update the functionality of the uC (realized in the software archi-tecture) and the FPGA (realized by the integrated memory circuit). The hardware design should allow for the detection of malfunctions of system components, e.g., a defect of the power output for the omni-directional camera.

Integration

The *Robot Head Control Module* should be located inside the robot head for an easy connection of all external units (e.g., motors). The mechanical design must be carried out under the consideration of the available space.

5.3.2   Software Architecture and Specification

The software architecture refers to the software components of the uC. Other unit components do not include software elements. Given the technical parameters of the determined uC, the architecture presented in Figure 5.12 is realized.

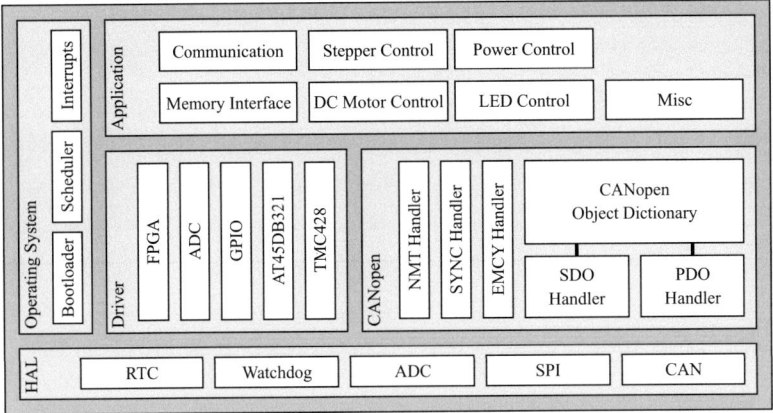

Figure 5.12: *Software architecture of the Robot Head Control Unit.*

Software Element Overview

Hardware Abstraction Layer (HAL)

This abstraction layer of the software architecture contains the interfaces to the hardware functionalities of the uC. Every component of the HAL includes registers for the configuration and usage of implemented hardware functionalities. For the aspired software system, the following components of the HAL are of relevance: The *Real Time Clock (RTC)* is required for the assessment of the current system time. This value is used for a synchronized control of the motors and the LEDs, and for the timing of outgoing CAN messages. The *Watchdog* is integrated to improve the reliability of this unit by restarting the uC in case of a software malfunction. The *ADC* is applicable for the determination of analog values of the camera power outputs, and the *SPI* and *CAN* modules are used for internal and external communication.

Operating System Layer

The operation system provides basic functionalities that can be used by all software components. This includes the *Scheduler* that processes the global system time of the unit and that calls software modules based on a given time flow. The *Interrupt Handler* includes functionalities required for the processing of internal and external interrupts (e.g., an incoming CAN message). The *Bootloader* module allows for an

115

update of all software components.

Driver Layer

Drivers provide software interfaces between the *Application Layer* and the *HAL* for configuration and information handling. The integration of drivers allows the usage of unmodified application modules for different hardware systems. For this unit, drivers are required that provide functionalities for the access of the FPGA, the Analog Digital Converter (ADC), the General Purpose Input/Outputs (GPIOs) as well as drivers for data exchange with the *Memory* and the *Stepper Controller*.

CANopen Layer

The CANopen protocol is used for the CAN based communication between the robot's control units. This protocol was developed for the application in automation under the direction of the company Bosch [Zeltwanger, 2008]. It provides services for network management, data transfer, and device monitoring. Of most relevance for the exchange of information with other units are the *Process Data Object (PDO)* and the *Service Data Object (SDO) Handlers*. PDOs are used to transfer real-time information. They include plenary process information. The CANopen protocol reserves ranges of CAN Identifiers (IDs) for the usage by PDO messages. SDOs are applicable for the transfer of configuration information or static process data. An SDO message includes information about the data type, the access type, and the address within the *CANopen Object Dictionary* of the transferred value. The *CANopen Object Dictionary* is the interface to the *Application Layer*. Within this dictionary, all values are defined (data type, access type, data range) that are accessible by the CANopen communication protocol. Data values of incoming SDO and PDO messages are stored in this dictionary; required information to be transmitted is taken from this dictionary.

Further services of the *CANopen Layer* are the *NMT Handler*, the *SYNC Handler*, and the *EMCY Handler*. The *NMT Handler* supports network management protocols that can be used to set a unit in different states (e.g., stop, start, reset) or for the generation of heartbeat messages determining the status of a unit. The *SYNC Handler* is applicable when several units need to process a task at the same time. Examples could be the synchronized control of actuators or the synchronized acquisition of process data. The *EMCY Handler* is responsible for the generation of CAN error messages in case of an internal error of a unit. This allows other units for the reaction to occurring malfunctions.

Application Layer

This layer represents all software modules that process the high level tasks of the unit. For the *Robot Head Control Unit*, this includes application modules for the control of the motors, the LED-matrix, and the control and monitoring of the camera power output. Other modules are responsible for the processing of incoming messages and the management of the FPGA configuration memory. The application module *Misc* includes functions, e.g., for the control of a unit status LED.

Non-Functional Requirements

System Update

The updatability of the functionalities of the uC and the FPGA must be enabled by the implemented software architecture. The upload of new functionalities must be possible over the CAN bus during normal operation. Incorrect update processes must be detected by the bootloader before the application is started. In such cases, the unit must enter a pre-operational mode and wait for a correct software update. The bootloader must always be accessible for the start of a new update process.

Software Timestamp

All PDOs must contain a timestamp for the validation of the actuality of the information. The global system time, transferred over the CAN bus, should be received by this unit. If the local unit time and the global system time are different, the local unit time has to be set.

Error Handling

In case of an internal error of the unit, a corresponding error message has to be sent using the CANopen *EMCY Handler*. Every error message must have a unique error code. Possible errors to be reported over the CAN bus could be a faulty communication to the FPGA, a short circuit or over-current on the power output of the omni-directional camera, or the detection of a blocked motor.

5.3.3 Unit Evaluation

The evaluation process of this unit must cover aspects of the hardware and software realization as well as the mechanical integration. For the verification of the hardware, test cases must be defined that cover all integrated modules regarding functionality and performance. Possible failure cases should be defined and tested, e.g., a reverse polarity of the power supply, or short circuits of the output signals. Finally, the overall power consumption should be measured and compared to the expected consumption of all components.

In a similar way, test cases have to be defined for the evaluation of all software functionalities. Additionally, the timings of all software modules should be measured and compared to the estimated values. The completeness and correctness of the *CANopen Object Dictionary* has to be checked. The generation of error message has to be proven as well as the correct functionality of the bootloader.

The evaluation process of the mechanical integration checks, whether this unit can be integrated into the robot head as planned. This realization has to be tested regarding producibility and maintainability (e.g., accessibility of connectors).

5.4 System Element Realization

In this project stage, all specified system units have to be designed. For embedded system units, this includes the physical design of the PCB under the consideration of mechanical constraints, and the writing of software programs. Every unit has to be tested regarding the specified evaluation procedures. In the course of this work, 16 units were developed to realize the embedded system of the shopping robot (see

Section 5.5). A total number of ten embedded system units were developed in accordance to the unit architecture described in the previous section. Three units are based on a simplified embedded system architecture (without CANopen stack).

Figure 5.13 presents the realized *Robot Head Control Unit* in the design stage (left panel) and the production stage (right panel). All required components, specified by the hardware architecture, were integrated. Mechanical requirements of the shape for the integration into the robot head were considered. The placement of connectors followed the rules for a good producability. The placement of other components was arranged by functional groups, which minimizes the impact of power components (e.g., motor drivers) to logical components (e.g., FPGA or uC).

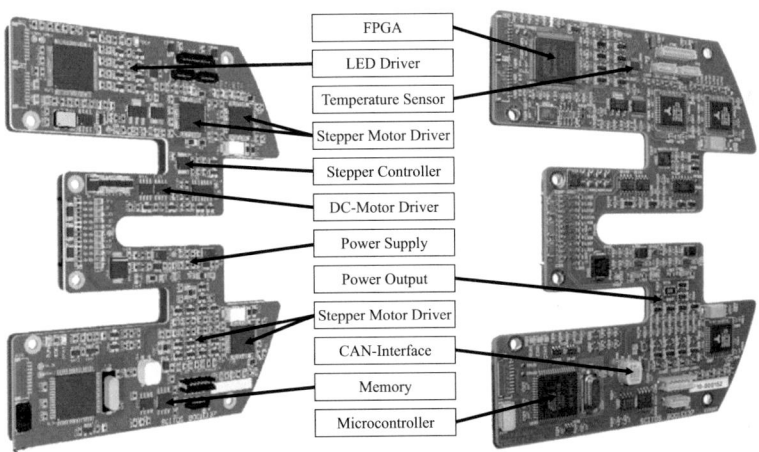

FPGA
LED Driver
Temperature Sensor
Stepper Motor Driver
Stepper Controller
DC-Motor Driver
Power Supply
Power Output
Stepper Motor Driver
CAN-Interface
Memory
Microcontroller

Figure 5.13: *Realization of the Robot Head Control Unit. The left picture shows the design state, the right picture the produced and assembled PCB.*

The implemented software was partitioned based on the software architecture. A re-usage of software components from other system units was possible for the modules of the *Operating System* and the *CANopen* layer, both developed at the beginning of the project. Drivers and application modules were specifically developed for this unit. The *CANopen Object Dictionary* was adapted regarding the transferred process information.

Table 5.10: *Power consumption of relevant components of the Robot Head Control Unit in accordance with the component datasheets.*

Hardware Element	Type	Consumption (@ 3.3 V)
Microcontroller	AT90CAN128	8.0 mA
FPGA (V_{CCIO})	EP1C3T100C8N	4.0 mA
FPGA (V_{CCINT})	EP1C3T100C8N	20.0 mA
Memory	AT45DB321D-SU	3 uA
CAN-Interface	SN65HVD233D	10.0 mA
Stepper Motor Controller	TMC428-I	1.0 mA
Stepper Motor Driver (3x)	TMC246A	2.2 mA
	Total	**45.2 mA**

The functional tests of this unit were carried out before this unit was integrated into its dedicated system segment. Therefore, all peripheral elements (motors, LED-matrix, camera) were connected to this unit and the expected functionalities and performances verified. For the electrical components, the overall power consumption was estimated based on the given manufacturer data (Table 5.10). This was compared to the real power consumption of the unit (measured value of about 39.0 mA @ 3.3 V). This further indicates the correctness of the physical design.

Table 5.11: *Measured time durations of relevant software components of the Robot Head Control Unit.*

Software Element	Layer	Period	Time
Scheduler	Operating System	10 ms	20 us
Interrupt Handler	Operating System	1 ms	50 us
Stepper Control	Application	50 ms	500 us
DC Motor Control	Application	50 ms	1,050 us
LED Control	Application	10 ms	3,600 us
Power Control	Application	10 ms	10 us
Communication	Application	10 ms	200 us
SDO Handler	CANopen	on request	250 us
PDO Handler	CANopen	10 ms	700 us

The functional tests of the unit were also applied to verify the correct functionality of the software. Furthermore, the correctness of the *CANopen Object Dictionary* was tested, because this component represents the interface to other system units. Finally, the operation times of certain software components were checked for plausibility (Table 5.11).

5.5 System Integration

The next step in the V-Model development process is the composition process of the robot system starting at the unit level. This process leads to the composition of functional groups at segment level (containing units), subsystem level (containing segments), and system level (containing subsystems). For every hierarchical level, the required evaluations have to be carried out. The composition process and the outcome of all evaluation processes must be documented. At the end of this project stage, the system is in deliverable form.

5.5.1 Unit Level

A total number of 38 system unit types composed the embedded system of the robot: 16 units, designed in the course of this work, and 22 external units (off-the-shelf products). The selection criteria for the external units followed functional and non-functional requirements and the decisions of the AHP. Examples are the *Laser Range Finder Unit* with an S300 laser scanner (company Sick) for a robust localization and navigation; the *Front Camera Unit* with a high-resolution camera for 3D obstacle detection and mapping [Einhorn et al., 2010]; the *Drive Motor Unit* containing Electronic Commutated (EC) motors with external rotors for a high torsional moment of the robot platform.

The 16 units, developed within this work, can be divided into three hardware units

(*Backplane Unit, Power Control Unit*, and the *Robot Head Light Unit*) and 13 embedded system units. These embedded system units have the following tasks:

Interface Unit: This unit controls the functionality of the embedded PC. It provides the power supply of the PC and monitors relevant signals of the motherboard. This gives the *Interface Unit* the possibility to turn on and off the embedded PC and to supervise its functionality.

Motor Control Unit: This unit is responsible for the control of both drive motors. It produces the power outputs for the current feed of the motor phases and calculates the robot's odometry based on the integrated hall sensors of the motors. For safety reasons, the collision sensor (bumper) of the robot and the optional emergency buttons are connected to this module to immediately stop the motors in case of a collision or user interference. This unit contains a uC and an FPGA for redundancy.

Sensor Control Unit: This unit primary controls the functionality of the *Ultrasonic Sensor Array Unit*. Therefore, the *Sensor Control Unit* contains an I2C interface for communication and a power supply output adapted to the requirements of the ultrasonic sensors. All incoming data packages from the ultrasonic sensors are checked for plausibility to detect faulty sensors already at this stage. This module further contains eight input/output ports for analog and digital signals.

Ultrasonic Sensor Array Unit: This sensor array represents multiple PCBs, each connected to one ultrasonic sensor. The size of this array can be adapted according to the requirements of the platform. For the shopping robot, 24 ultrasonic senor modules are combined to build the *Ultrasonic Sensor Array Unit*.

EBC Unit: The *External Bus Connector Unit* provides two EBC ports for the connection of units (e.g., the *Robot Head Control Unit*). Each EBC port contains a power output for the battery voltage (max. current of 4.0 A) and power outputs for 5.0 V and 12.0 V (max. current of 2.5 A, each). Every power output can be separately turned on and off, and is monitored regarding over-current. The over-current threshold leading to an emergency shutdown of a power output can be configured by software. Additionally, every EBC port includes a CAN interface.

Robot Head Control Unit: The *Robot Head Control Unit* is responsible for the control of all components of the robot head (as described in the previous sections).

Head Camera Unit: This unit belongs to the omni-directional camera of the robot. It contains a USB hub for the connection of four cameras. A small uC can be used to turn on and off these cameras.

Display Control Unit: The *Display Control Unit* controls all functionalities of the *Display/Multimedia Segment*. This includes the power supply of the display and the touch sensor as well as the power supply and data exchange of the *Barcode-Reader Unit*. Furthermore, a USB sound card is integrated to generate sound outputs (two loudspeakers) and to receive speech input (four microphones).

Battery Control Unit: This module is responsible for the management of the robot's battery. The unit's tasks include the monitoring of all cell voltages, the balancing of the cells, the control of the charging process (setting of current and voltage levels), and the system shut down in case of failures (e.g., under-voltage, over-voltage, or over-current). Because this module is directly connected to the battery, it has further responsibilities for the control of the whole robot system. These responsibilities are: the execution of the power-up procedure of the robot, the generation of the global system time, the execution of a determined power-down procedure, and the wake-up of all system components after a defined sleeping period or in case of a user request.

The technological progress in battery technologies during the time course of this work led to the development and implementation of two types of battery systems. Therefore, two types of *Battery Control Units* were developed. The first version (developed during the SerRoKon projects) controls two lead-acid battery cells (nominal voltage of 24.0 V and nominal capacity of 42.0 Ah). This *Battery Control Unit* includes a 100 W line voltage power converter for the manual charging of the battery. The second *Battery Control Unit* (designed in an additional development cycle after the finishing of the SerRoKon projects) controls the 56 cells of a lithium-polymer battery (nominal voltage of 25.9 V and nominal capacity of 80.0 Ah). This version is able to handle seven lithium-polymer cells in series and is integrated inside the casing of the battery (Figure 5.14). For the manual charging process based on line voltage, an external *Power Supply Unit* with 500 W is used.

Both battery versions were positioned at a very low and central point of the robot platform (Figure 5.15). This guarantees the very high stability of the shopping robot system.

Auxiliary Charger Control Unit: In addition to the manual charging mode based on line voltage, the AHP proposed the realization of an extra-low voltage autonomous charging mode. Therefore, the *Auxiliary Charger Control Unit* was designed, which converts the incoming extra-low voltage from the autonomous charging station to voltage and current levels suitable for the charging process of the robot's battery. This unit communicates with the autonomous charging station to start the charging process. It further controls

Figure 5.14: *Lithium-polymer battery of the shopping robot platform without cover. The Battery Control Unit connects the integrated 56 battery cells.*

Figure 5.15: *Shopping robot base with the drive system and two lead-acid batteries.*

two motors that push the charging contacts of the robot downwards (to contact the metal plates of the charging station).

Autonomous Charger Control Unit: This unit is the counterpart of the *Auxiliary Charger Control Unit*. It is installed inside the autonomous charging station and enables the power output to the charging metal plates.

User-RFID-Reader Unit: The *User-RFID-Reader Unit* was developed to enable operators to set the robot applications into special operating modes (e.g., an administration mode). This unit contains an RFID reader with a short detection range. If an RFID tag is placed close to this module, a CAN message containing the detected RFID code is generated.

Status Display Unit: This optional module can be installed on the robot. It contains a small display and a turning knob, which allow for the configuration of low level system parameters (e.g., the control of the ultrasonic sensors, or the EBC ports).

5.5.2   Segment Level and Subsystem Level

The developed hardware units, embedded system units, and the selected external units were composed on segment and subsystem level. Figure 5.16 presents an

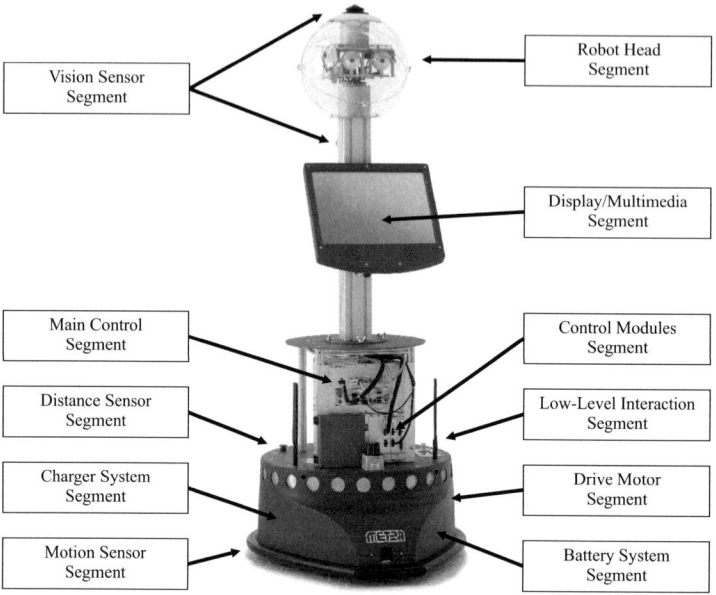

Vision Sensor Segment		Robot Head Segment
		Display/Multimedia Segment
Main Control Segment		Control Modules Segment
Distance Sensor Segment		Low-Level Interaction Segment
Charger System Segment		Drive Motor Segment
Motion Sensor Segment		Battery System Segment

Figure 5.16: *System segments of the finished SCITOS G5 platform.*

overview of the robot platform and the integrated segments.

5.5.3   System Level

The final step of the composition process is the combination of the subsystems to the overall robot system. At this stage, the interaction of all system elements can be evaluated.

The system architecture recommended by the AHP (Figure 4.5) was realized for the embedded system design. The realized system architecture of the shopping robot including all internal communication interfaces is presented in Figure 5.17.

The redundancy of this modular system architecture leads to the high safeness and robustness of the shopping robot system. Both criteria were highly prioritized by the AHP. Some of the supported safety features are:

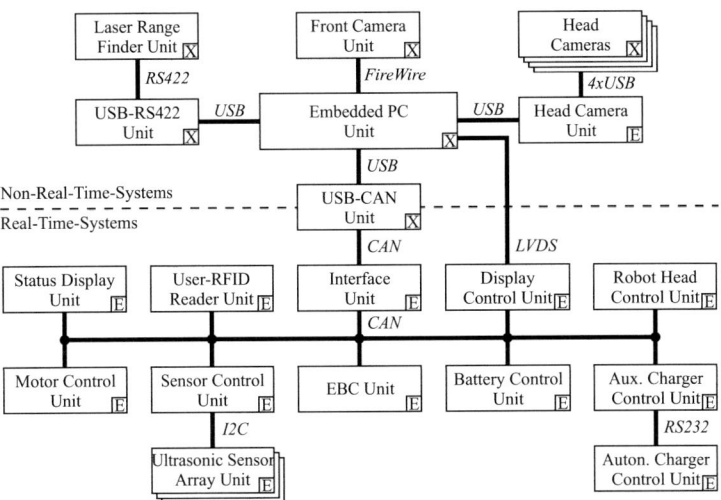

Figure 5.17: *System architecture of the shopping robot system including Embedded System Units (E), External Units (X), and all communication interfaces.*

- Control of the functionality of all modules based on periodical heartbeat messages.

- Usage of time stamps to prove the functionality of the communication system.

- Control of the functionality of the *Embedded PC Unit* based on status flags and the power consumption.

- Implementation of watchdog functionalities, e.g., to stop the motors after a given time without data update.

- Hierarchical over-current detection and shut-off.

- Redundancy in the monitoring of safety sensors.

- Low-level communication of system events, e.g., unit failures.

The detection of a system failure causes a system reaction depending on the seriousness of the failure. For example, a battery under-voltage causes a complete system shutdown; missing driving information causes the stop of the motors; a short-circuit in the *Robot Head Segment* causes the shut-down of this segment without an influence on other functionalities. The implemented safety concept was verified by the German Technical Inspection Agency (TÜV).

Functional tests at this project stage were arranged in the development and manufacturer laboratories. In the first instance, these tests included the verification of all safety features, the reliability of all system functionalities (drive system, charging systems, communication systems), and the evaluation of the aspired system parameters (maximum speed, payload, power consumption, runtime).

5.6 System Delivery and Evaluation

5.6.1 Field Tests

In the final stage of the system development process, the fulfillment of all functional and non-functional requirements has to be verified. For the shopping robot platform, these evaluation processes were arranged in home improvement stores of the project partner Toom BauMarkt GmbH. The tests were carried out in two periods. In the first step, three robot systems were installed in a store with a size of about $7,000\,m^2$. The robots were required to bring customers to requested product places (see Section 5.1.1). In a test period of five months (April to August 2008), the robots drove about 600 km and were in dialog with 3,764 users [Gross et al., 2009].

In a second test period, two additional stores were equipped with three robot systems each. These stores had a size of $8,000\,m^2$ and $10,000\,m^2$, respectively. In a period of four months (November 2008 to February 2009), the nine robots in all three stores served 9,515 customers. The overall driving distance of all robots was about 1,600 km. A survey at the end of a guided tour revealed that 87.9 % of the users were satisfied with the functionality of the robot and 88.5 % would use the robot again. About one fifth of the users, successfully finishing the tour, participated in the survey.

The two test periods showed that the robot systems are able to satisfy walk-in users and to robustly move through large store environments. The main problems identified during these tests were blocked ways, which prevent the robot from reaching its goal or the loss of localization. Improvements of these constraints were carried out in additional development steps that were not in the focus of this work.

The test periods further revealed two technical problems of the robot system. First, components of the *Motor Control Unit* were aging very fast, because of the permanent stress in the shopping application. This led to a worsen driving behavior or even the fall out of the drive system. Second, the high temperature of the charging station during the charging process led to the fall out of the integrated power converter in some cases. The execution of the long-term field tests was necessary to find these design failures. Both problems were corrected after the field test period.

5.6.2 Satisfaction of the AHP Criteria and Evaluation of the Development

This section encompasses the fulfillment of the development criteria weighted for the shopping robot application. The determined impact of each criterion (Section 5.1.3) is given in brackets:

Adaptability (6.7 %): The development process did not prioritize the adaptability of the robot system. However, the realization of the modular system architecture allows a highly flexible adaptation of the robot system to new application areas (see Section 5.6.3). Moreover, the integration of a lithium-polymer battery further supports novel applications.

Operation Time (4.2 %): The overall power consumption of the robot system is less than 100 W (in average), because of the integration of consumption optimized components and the possibility to turn-off unused system units. Based on the 42 Ah lead-acid battery and the developed charging system, the robot system can operate up to eight hours with one battery charge and requires the same time to re-charge its battery. These values are sufficient for the intended usage.

The implementation of the 80 Ah lithium-polymer battery in an additional development cycle allows for operation times of up to 16 hours. The charging time (for an empty 80 Ah battery) is less than eight hours. This battery system completely satisfies this criterion.

Usability (17.0 %): The field tests prove the usability of the robot system. The integration of system elements for user interaction (i.e., display, loudspeakers, and microphones) was suitable. The size and shape of the robot as well as the realization of the robot head appear to be highly acceptable by users.

Robustness (14.6 %): The design of system units considered the moving character of the robot platform. For example, bigger PCBs components are glued to avoid drop offs; the usage of aging components (e.g., electrolytic capacitor) was avoided if possible. For the selection of external units (e.g., laser range finder or the embedded PC), industrial components were preferred over consumer products.

Safeness (37.4 %): *Safeness* was the most important criterion for the development process, which was addressed by system redundancy and the integration of additional security functions (e.g., forced reduced driving speeds in narrow areas). The passing of all evaluation processes of the German Technical Inspection Agency (TÜV) demonstrated the complete satisfaction of this criterion.

Features (6.7 %): The integrated technologies are limited to the requirements of the shopping robot application. A special attention to additional feature for future developments was not arranged. Nevertheless, the high adaptability of this robot allows for the integration of additional features (e.g., depth cameras).

Costs (13.4 %): The development process was carried out under consideration of costs aspect. However, the integration of a modular system architecture and industrial components, required for the satisfaction of the criteria *Safeness* and *Robustness*, forced higher system costs than expected. The design of a mobile robot system for public applications is a trade-off between required safeness and acceptable costs. The successful consideration of both aspects can be determined on the sales figures of a service robot (see Section 5.6.3).

5.6.3 Application of the Shopping Robot System

In addition to the described application in home-improvement stores, the shopping robot system is used in other shopping and guidance applications. Examples are the two innovation guides *Roger* and *Ally* in the *real,- future store* in Tönisvorst (Figure 5.18, top-left), or the shopping guide *Werner* in several *Conrad electronic* stores in Germany. The developed system was also applied to exhibitions and trade fairs, e.g., Macropak 2010 in the Netherlands or Ecomondo 2011 in Italy (Figure

Figure 5.18: *Further applications of the shopping robot system. Top-Left: innovation guide Ally in the real,- future store in Tönisvorst [Metro, 2011]. Top-Right: monitoring robots in the production clean room at Infineon Technologies in Dresden [Infineon, 2011]. Bottom-Left: a SCITOS A5 guidance system at a trade fair in Italy. Bottom-Right: a SCITOS G5 platform used for the research on RFID-based localization algorithms [Zell, 2011].*

5.18, bottom-left).

Moreover, the high adaptability and the robustness of the developed robot system allows for the usage of this system in other application areas. Successful realizations in the industrial sector are the monitoring robots at *Infineon Technologies*. These robots are equipped with analysis devices to control the air contamination in clean room production facilities (Figure 5.18, top-right). The robots drive autonomously to designated locations to check environmental parameters based on a programmed measuring plan.

In the research sector, the robot base is used for the development of algorithms for human-machine-interaction [Kraft et al., 2010], localization (Figure 5.18, bottom-right) [Vorst et al., 2011], or navigation [Nehmzow, 2009]. For the research on manipulation tasks, the robot platform is also used in combination with robot arms that are connected to the robot's power supplies and communication interfaces [Burbridge et al., 2011].

Chapter 6

Development of the Home-Care Robot Platform

In the previous chapter, a system design method for complex robot systems was engineered on the example of a shopping robot application. The V-Model was adapted for the specific demands of a robot system. Interestingly, the AHP was used to support the decisions in the V-Model process. This required the careful selection of decision criteria and the weighting of design alternatives.

In this chapter, the principles of the developed system design method will be generalized and verified on a second service robot system for home-care applications. This robot system, developed in the course of the *CompanionAble* project (see Chapter 1.1), is planned to help elderly persons staying longer in their familiar environments. This chapter will demonstrate that, although the shopping and the home-care application have different system requirements, operation areas, and functionalities, the suggested system design method (V-Model + AHP) can be successfully applied.

6.1 System Specification

Initial Situation and Objectives

The number of persons with cognitive disabilities rises with the increasing average life expectancy. Especially, people with Alzheimers disease require a comprehensive care to avoid isolation from the society. In many cases, family members are over-strained with the required nursing effort or do not have the possibility to care for an elderly person. The inclusion of nursing staff is restricted to time-constrained tasks several times per day. Therefore, it is often required that Care Recipients (CRs) move to nursing homes, which usually reduces the life quality of the CR.

The usage of assistive service robots could postpone the date, when a CR must leave his/her home environment to move to a nursing home. The robot system could take over important tasks that assist a safe living of the CR in his/her home environment. Such tasks might include the monitoring of the health status of a person (e.g., behavior), the detection of fall, or the reminding of actions (e.g., taking medicine or storing keys in the robot's tray). An assistive robot system could further be used to enhance the mental activity of a person or it could function as a communication device for the interaction with family members, friends, or the Care Givers (CGs) [Gross et al., 2011a], [Gross et al., 2011b].

The goal of the development described in the following sections is the design of an interactive service robot platform that provides adequate technologies for the fulfillment of the described tasks. The robot system must be able to navigate in home environments and to interact with elderly persons. It must be able to detect critical situations of the CR based on the integrated sensor systems and to alert CGs, if necessary. The robot should support family members and CG based on interaction and monitoring functionalities. It is not intended to manipulate objects or persons.

Functional Requirements

The following use cases cover the main functionalities that need to be considered during the robot platform development process:

Use Cases 1 : Searching for a Person

Abstract: The robot drives autonomously through the home environment and searches for a person in any position (standing, sitting, or lying).

Sequence:

- The robot drives through the home environment and avoids collisions with obstacles.

- Using video input, sound input, and the input from a depth camera, the robot tries to detect a person.

- If the robot detects a person, it approaches the person using its navigation sensors.

- After the robot reaches the assumed position of the person, it starts interaction.

Subtasks:

- Navigation through a given environment

- Obstacle detection and collision avoidance

- Drive deactivation in case of a collision

- Detection of persons in the given environment

- Drive to the position of the person and approach the person

- Sound replay to interact with the person

- Input of information on a touch screen

- Input of speech commands by microphones

Use Cases 2 : Interaction with a Person

Abstract: The robot drives on request to a person and interacts with this person.

Sequence:

- The person calls for the robot.

- The robot detects the call and identifies the command.

- The robot detects the position of the person and drives to this location.

- After reaching the position of the person, the robot offers its services based on a touch screen and speech output.

- The person selects a service based on voice control or by using the touch screen.

- The robot starts, e.g., a video telephony to a family member.

Subtasks:

- Identification and analysis of voice commands

- Estimation of the direction, where the voice command was coming from

- Driving to a position and searching for a person

- Interacting with a person based on voice input and speech output

- Interacting with a person based on a touch screen

- Execution of video telephony

Use Cases 3 : Remote Control

Abstract: The robot is remotely controlled by a family member or a Care Giver (CG). It avoids the collision with obstacles.

Sequence:

- A CG controls the robot by means of a wireless connection.

- The CG is able to see a video of the robot's surrounding.

- The CG is able to control the movement of the robot.

- The robot detects obstacles and stops the movement before it collides.

Subtasks:

- Connection of a CG to the robot based on a wireless connection

- Control of the drive system based on a wireless connection

- Transfer of video information of the robot's surrounding

- Monitoring of the remote control process and stopping of the motors before a collision occurs

Use Cases 4 : Charging Process

Abstract: The robot diagnoses an empty battery, drives to a charging station, and docks-on to the this station.

Sequence:

- The robot detects that the battery charge drops below a defined threshold.

- The robot stops its current task and drives autonomously to the location of a charging station.

- The robot detects the charging station and starts the docking process to this station.

- After reaching the final position, the robot starts the charging process.

- In case of a user request or refilled batteries, the robot docks-off from the charging station.

Subtasks:

- Estimation of the robot's battery charge

- Driving to a given location, avoidance of collisions

- Detection of a charging station and docking to this station

- Starting and stopping of the charging process

Requirement Lists

The following requirements lists were compiled based on the described use cases and classified into functional groups. Every requirement is weighted by a priority (mandatory (M), desirable (D), or optional (O)).

Table 6.1: *Mobility requirements*

Requirement		Prio.	Description
1.1	Stability	M	Stable movement and standing characteristics provided by the mechanical framework and the drive system.
1.2	Motor controller	M	Speed control by the integrated motor controller. Stop of movements in case of a collision or failure.
1.3	Accurate position	M	Accurate position and movement information of the robot provided by integrated odometry sensors.
1.4	Avoiding of collisions	M	Equipment of the robot with a set of sensors to detect obstacles in the environment and to avoid collisions.
1.5	Detection of collisions	M	Detection of collisions by an integrated security sensor.
1.6	Operation Time	M	Operation of the robot system with a minimum of ten hours with one battery charge.
1.7	Charging Time	O	Charging an empty battery in less than five hours with a running PC.
1.8	Autonomous charging	M	Recharging of the robot by an autonomous charging system without user interference.

Table 6.2: *Functionality requirements*

Requirement		Prio.	Description
2.1	Battery charge	M	Estimation of the battery charging state.
2.2	Robot charging	M	Charging by an integrated charging system connected to a standard power plug.

2.3	Charging station	M	Usage of an autonomous charging station to avoid operator interference.
2.4	Pleasant appearance	M	The size and shape of the robot must be pleasant for users in home environments.
2.5	Identification of persons	M	Detection and identification of persons in standing, sitting, and lying positions.
2.6	Switchable power supplies	D	Software used to switch on and off the power supply of integrated modules.

Table 6.3: Usability requirements

Requirement		Prio.	Description
3.1	I/O by touch screen	M	Integration of a touch screen to show information or to modify the robot behavior by touch input.
3.2	Sound/Speech output	M	Sound output by a set of loudspeakers.
3.3	Directional speech input	M	Voice input by directional microphones for command processing and video conferencing.
3.4	Robot head	M	Integration of a robotic head to attract users, show basic emotions, and indicate the robot status.
3.5	WLAN interface	M	Exchange of information based on wireless communication.
3.6	RFID tray	O	Integration of an RFID reader into the robot's tray to identify objects (e.g., keys or wallets).

Table 6.4: *Service requirements*

Requirement		Prio.	Description
4.1	Internal errors	M	Detection of internal errors and communication of errors to a service point.
4.2	Installation	M	A simple installation process, which could be carried out by skilled private persons.
4.3	System analysis	O	Analysis of the system state by remote access.
4.4	Transportation	D	Availability of a transportation box to protect the robot during shipping.

Non-Functional Requirements

Working Area

The home-care robot system has to be designed for home environments. The project partners agreed that the robot's operation area does not include sleeping rooms and bathrooms for privacy reasons. Therefore, comparable environment characteristics (temperature range, humidity, pollution) as in the shopping application can be assumed. The questioning of CRs has shown that in most flats (73 %) all doors are wider than 72 cm, which determines the maximum size of the robot. Furthermore, typical floors are made from wood, parquet, carpets, or tiled floor. Step heights are usually smaller than 15 mm (two-third of the interviewees). To simplify the development process of the robot platform during the research project, it is also assumed that the given surroundings are free of ledges, where to robot could fall down.

The working areas in home environments are much smaller than in stores or exhibitions. For example, the test environment, which is provided by a project partner, has a living space of about $150\,m^2$ (Figure 6.1). The robot operates in two-thirds of this area. Other apartments are assumed to have a similar area.

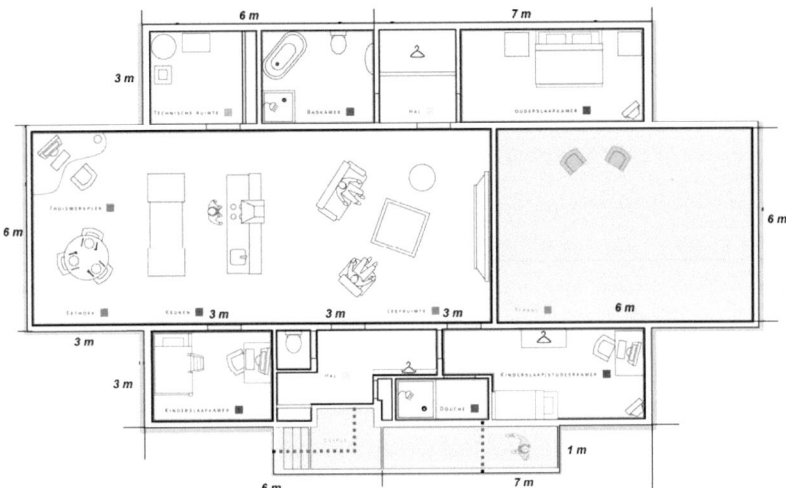

Figure 6.1: *Map of the test environment at the project partner SmartHomes in Eindhoven (Netherlands) [Companionable, 2011].*

Another aspect in home environments is the existence of obstacles in various heights that cannot be seen by a laser range finder, placed in the lower part of a robot platform. Examples are tables, chairs, or opened cabinet doors. The robot platform sensor configuration must allow for the detection of such obstacles. Furthermore, navigation and localization algorithms have to be developed under the consideration of the changing positions of obstacles in home environments (e.g., chairs).

Constraints of the intended robot system are closed doors in the flat. This robot platform will not be able to open doors by physical manipulation (e.g., based on a robot arm). The usage of the robot is restricted to areas, where it the can freely move. Another possibility would be the integration of automatic doors, which might be applicable for nursing homes or hospitals.

Physical Constraints

The design process of the robot has to consider the parameters of the home environments. This mainly covers the minimum width of doors and expected barriers heights. For an easy passing of doors, the width of the robot platform should be significantly smaller than the door width, because of inaccuracies of integrated sensors. A robot width of less than 50 cm seems to be applicable. To pass barriers (e.g., doorsteps or carpet edges), the robot platform should be able to drive over barrier heights of at least 1.5 cm.

The design of the robot must account for a high acceptance by users. The height of the robot must avoid a frightening appearance, which might be the case, if the robot is much higher than, e.g., a sitting person. Still, the robot must be usable in standing and sitting positions. An overall height of the home-care robot of about 1.2 m seems to be a good compromise between both requirements. Additionally, the robot design should have a friendly appearance. Technical details (e.g., integration of sensors) should be covered by the design of the casing as good as possible. A good example for the realization of a social robot hiding technical details is the seal robot *Paro* [Wada and Shibata, 2008].

Reliability and Operation Time

In an optimal case, observation and interaction tasks would require the continuous operation of the robot system. Such a continuous operation cannot be realized by systems based on batteries. To account for this problem, the operation time of the robot system should be optimized by increasing the battery capacity, decreasing the power consumption of system components, and by shortening of the charging process. A smart management of the time, when the robot could re-charge its batteries, e.g., when the CR is sleeping or not at home, would be advantageous. Further, during charging, the robot could remain in an active state, where it is still able to detect emergency situations or react to calls. Working times with one battery charge of more than ten hours and a charging time (for an empty battery with all system components activated) of less than four hours would be reasonable for the planned application.

The failure rate of the system should also be as low as possible. A down time of less than three days per year would be preferable. Special attention should be given during the development process to avoid a total fade-out of the robot, so that the robot is able to communicate its current state to the user.

Installation and Service

Instructed service personal should be able to carry out the installation process of the robot system, including the mapping of the home environment and the adaptation of the application to given constraints. The assistance of elderly people require that the robot reacts to system failures in an adequate ways (speech output to explain the current situation) and to communicate this event to a service point. Service personal must be able to check the correct functionality of the robot by remote access.

Overall System Architecture and Life Cycle Analysis

Overall System Architecture

Figure 6.2 presents the concept of the overall system architecture. Similar to the shopping robot system, it consists of embedded control systems, power supply systems, sensor systems, the drive system, and interaction systems.

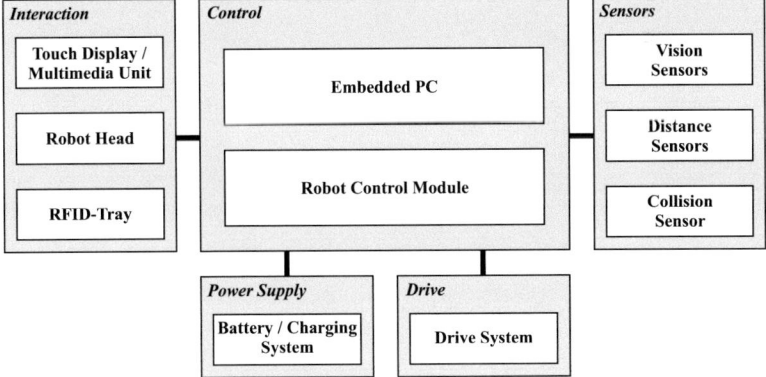

Figure 6.2: *Overall system architecture of the home-care robot platform. This figure represents a functional view of architecture, independent of the later technical realization.*

Interface Overview

The human-robot interfaces are mainly realized by the integrated touch screen, microphones, loudspeakers, and the robot head including the robot's eyes and a pet sensor. Another system, belonging to the human-robot interface is an RFID tray, which can be used to store important objects. The robot can be turned off by a toggle switch, which is useful for the transportation of the system. The WLAN interface should allow family members or CGs to communicate with the CR, or an operator should be able to monitor the system state.

The robot's WLAN should also be as a supporting system interface. In this case, the robot could receive information (e.g., news, weather forecasts) from service provider. The communication to the charging station also belongs to the supporting system interface.

System Life Cycle

Similar to the shopping robot system, technological progress (i.e., battery, sensor, and computing systems) require a periodical revision of the robot's technical systems. The expected life cycle of the home-care robot system is also about three years.

6.1.1 Specification of the Evaluation Process

The evaluation process should cover all use cases determined by the functional requirements. The test procedures should be carried out at the developer's and user's side. In the laboratories at developer's side, parts of a home environment are arranged to execute functional testing (e.g., approaching a person). The tests at user's side should be carried out in a reference apartment (Figure 6.1) of a project partner, in which participants can life for a given time. This test environment should fulfill all requirements of an intended home environment (e.g., small barriers, sliding doors).

6.1.2 Evaluation of AHP Criteria for the Home-Care Robot Application

Based on the described functional and non-functional requirements, the AHP criteria can be evaluated:

Adaptability (A): The robot system will be exclusively developed for assistive applications in home environments. The application in other working areas (e.g., hospitals) will be arranged based on the given system realization. An adaptation of the robot system (embedded system and mechanical design) to other fields of applications is not planned. Therefore, this criterion is of very low priority.

Operation Time (O): For the best usability, the robot must move freely in the home environment to monitor for CR or to approach to the CR for interaction purposes. During the charging process, which should be carried out autonomously, the functionality of the robot is restricted to stationary tasks (e.g., stationary user monitoring, waiting for incoming commands). Even if the robot still provides a set of functionalities in the charging mode, the operation time with one battery charge should be maximized and the charging time minimized. The satisfaction of the criteria *Operation Time* is weighted to be "demonstrated" more important than *Adaptability*.

Usability (U): This criterion is of high relevance for the intended robot application. Particularily, the interaction with CR suffering from cognitive impairments requires an intuitive and easy usability of the robot system. The design process has to consider the integration of all interactive components in an adequate way (i.e., the position of the display and the understandability of speech outputs) and a technology-hiding integration of all system components. This criterion is weighted to be "demonstrated" more important than *Adaptability* and weakly more important than *Operation Time*.

Robustness (R): The robustness of this robot system is of relevance for the success of this application. Nevertheless, it is expected that the mechanical stress is lower than in shopping applications, because of the homely working area and the lower driving speed. This criterion is, therefore, weighted to be "demonstrated" more important than *Adaptability*, equally important as *Operation Time*, and weakly less important than *Usability*.

Safeness (S): The safeness of users and home furnishings has to be considered during the development process. The diversity of obstacles that could be hit or knock over by the robot has to be considered. The realization of a safe charging system is also of relevance for design decisions. The criterion *Safeness* is weighted as "demonstrated" more important than *Adaptability*, equally important as *Operation Time*, weakly less important than *Usability*, and weakly more important than *Robustness*.

Features (F): The integration of features is of relevance for the later implementation of novel functionalities (e.g., enhanced person tracking algorithms) based on the same robot platform. For the weighting process, it is assumed that this criterion is weakly less important than *Operation Time* and *Robustness*, and essentially less important than *Usability* and *Safeness*. This criterion is weighted to be essentially more important than *Adaptability*.

Costs (C): The production and maintenance costs are important parameters for the market penetration and the success of this robot platform. The goal of the development process would be to achieve a price that allows for the marketing of this platform in the consumer sector. Project partners agreed that a selling price of less than 10.000 Euro would be desirable. Consequently, *Costs* are absolutely more important than *Adaptability*, strongly favored over *Features*, absolutely more important than *Robustness*, and weakly more important than *Operation Time*, *Usability*, and *Safeness*.

The results of the pairwise criteria comparisons and the calculated weights are presented in Table 6.5.

Table 6.5: *Criteria evaluation matrix of the home-care robot platform. The structure of this table is in accordance to Table 5.5.*

	A	O	U	R	S	F	C	Weights	C.R.
A	1/1	1/7	1/7	1/7	1/7	1/5	1/9	**2.1 %**	
O	7/1	1/1	1/3	1/1	1/1	3/1	1/3	**11.2 %**	
U	7/1	3/1	1/1	3/1	3/1	5/1	1/3	**22.6 %**	
R	7/1	1/1	1/3	1/1	1/3	3/1	1/5	**9.3 %**	0.067
S	7/1	1/1	1/3	3/1	1/1	5/1	1/3	**14.6 %**	
F	5/1	1/3	1/5	1/3	1/5	1/1	1/6	**5.0 %**	
C	9/1	3/1	3/1	5/1	3/1	6/1	1/1	**35.2 %**	

In comparison to the shopping robot application, the results of this weighting process show that the development must strongly focus on *Costs*, because of the usage of this platform in the private sector. The criterion *Safeness*, most important for the shopping system, is still of a remarkable interest, but less important than *Usability*. The criteria *Adaptability* and *Features*, which might be considered for further usage in other applications, are not of relevance for the development process. Figure 6.3 presents the criteria chart of the weighting results for the home-care robot system.

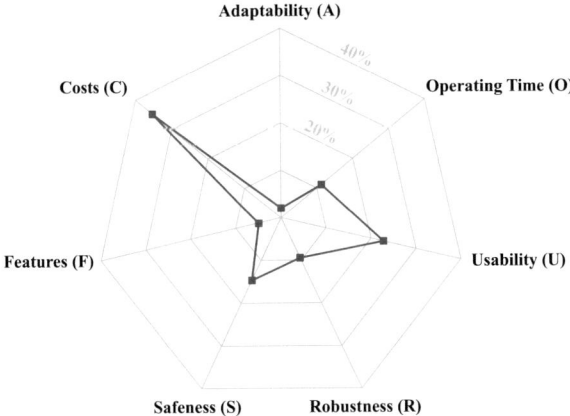

Figure 6.3: *Criteria chart for the home-care robot platforms representing the calculated criteria weights of Table 6.5.*

6.2   System Design

6.2.1   System Architecture

Given the criteria weights for the home-care robot system, an appropriate system architecture can be derived. For the described alternatives of the system architecture (Chapter 4.2.2), the AHP revealed the priorities shown in Table 6.6.

Table **6.6:** Decision results for the system architectures.

	A	O	U	R	S	F	C	Prio.	Rank
	2.1%	11.2%	22.6%	9.3%	14.6%	5.0%	35.2%		
A1	12.5%	36.4%	25.0%	5.4%	5.7%	8.3%	43.4%	**27.0 %**	3
A2	12.5%	36.4%	25.0%	14.6%	26.3%	8.3%	43.4%	**30.9 %**	1
A3	12.5%	6.6%	25.0%	23.7%	12.2%	41.7%	4.0%	**14.1 %**	4
A4	62.5%	20.7%	25.0%	56.3%	55.8%	41.7%	9.2%	**27.9 %**	2

In contrast to the modular architecture A4 that was used for the shopping robot system, architecture A2 was determined as most applicable for the home-care application. This system architecture consists of a centralized control unit (embedded PC) to execute complex software algorithms and a system-specific co-controller to process real-time tasks. The selection of this architecture is mainly driven by the criteria *Costs* and *Safeness*. The production costs of this architecture are low, because of the small number of embedded system units (primarily the co-controller) and the high specialization of this co-controller to the required functionalities. The high safeness of this system architecture is also enabled by the co-controller. An adequate design of this co-controller allows redundancy in the monitoring of safety functionalities and determined reaction times to critical system events.

The concept of the selected architecture is the processing of all low-level functionalities by one embedded system unit, the co-controller. However, the wide distribution of functionalities inside the robot system (e.g., the head) would require a complex cabling of these components to the co-controller. This increases the productions

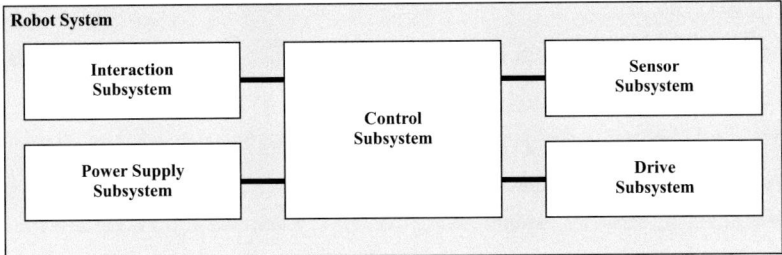

Figure 6.4: *Decomposition of the home-care robot system into subsystems.*

costs and decreases the robustness of the system. To address this problem, two additional low-level system units are designated: a head unit to control the multimedia systems and the robot head components (i.e., the eye displays); and a battery control unit to monitor, charge, and protect the robot's battery.

6.2.2 System Decomposition

Decomposition of the System into Subsystems

Equivalent to the shopping robot system, this robot consists of a *Control Subsystem*, a *Power Supply Subsystem*, a *Drive Subsystem*, a *Sensor Subsystem*, and an *Interaction Subsystem* (Figure 6.4).

Decomposition of Subsystems to Segments

The decomposition process from subsystem to segment level still leads to a system structure (Figure 6.5) similar to the shopping robot system. The main difference is the reduced number of segments, which was achieved by allocating several functionalities to few segments. For example, the *Main Control Segment* is responsible for the execution of complex software algorithms and for the processing of real-time tasks. Other segments with an increased number of functionalities are the *Head/Multimedia Segment* and the *Battery/Charger System Segment*. All three segments are described in more detail in the following decomposition step.

The remaining system segments have a similar functionality as described for the shopping robot: The *Motion Sensor Segment* includes the collision sensor; and the *Vision Sensor Segment* involves a front-camera and a camera at the backside of the robot supporting the docking process to the charging station. The *Distance Sensor Segment* includes a laser range finder and ultrasonic sensors. For the detection of higher obstacles, a depth camera is added to this segment. The *Low-Level Interaction Segment* is composed of an RFID tray, small objects can be stored.

The realization of the *Drive Motor Segment* was determined by the AHP. The results of the weighted drive system alternatives evaluated by criteria weighted for the home-care robot system are presented in Table 6.7.

Given these priorities, the home-care robot will also be based on a differential drive system with two driven wheels and one castor wheel (D4). Similar to the shopping robot, the driven wheels will be placed out of the center line of the robot platform. A detailed description of the realization of the *Drive Motor Segment* is presented in Section 6.5.

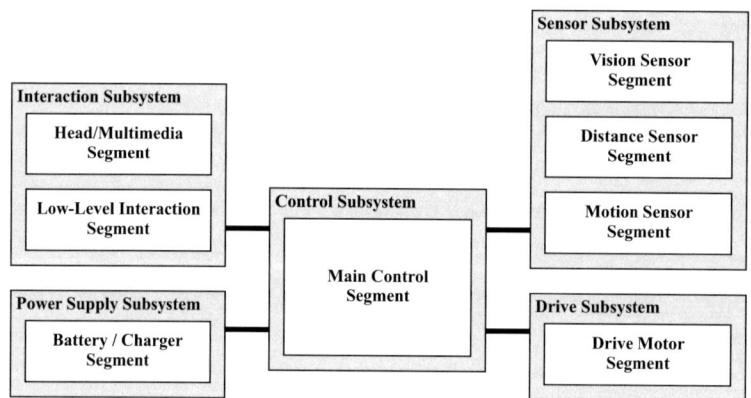

Figure 6.5: *Decomposition of the home-care robot from subsystem to segment level.*

Table 6.7: Decision results for the drive systems.

	A 2.1%	O 11.2%	U 22.6%	R 9.3%	S 14.6%	F 5.0%	C 35.2%	Prio.	Rank
D1	16.7%	3.9%	27.0%	42.0%	0.0%	16.7%	19.9%	**18.6 %**	3
D2	16.7%	19.2%	27.0%	19.2%	9.5%	16.7%	19.9%	**19.6 %**	2
D3	16.7%	19.2%	27.0%	8.0%	19.0%	16.7%	8.3%	**15.9 %**	4
D4	16.7%	19.2%	12.1%	19.2%	16.1%	16.7%	38.4%	**23.7 %**	1
D5	16.7%	19.2%	3.5%	8.0%	25.5%	16.7%	9.2%	**11.8 %**	5
D6	16.7%	19.2%	3.5%	3.7%	29.9%	16.7%	4.2%	**10.3 %**	6

Decomposition of Segments into Units

The decomposition process of segments into units is discussed on the three segments that were most relevant for this work. The decomposition results of the other segments can be found in Appendix C.

Main Control Segment

In accordance with the system architecture, the *Main Control Segment* contains the *Embedded PC Unit* and the robot co-controller in form of the *Main Control Unit*. The *Embedded PC Unit* is responsible for the computation of complex software algorithms (e.g., navigation, localization, or interaction) and for the storage of complex information (e.g., navigation maps). The functionalities of this unit are similar to the embedded PC of the shopping robot [Gross et al., 2011a].

The *Main Control Unit* executes all tasks that are carried out by the *Control Modules Segment* of the shopping robot system. This unit is responsible for the control and monitoring of the *Embedded PC Unit*, the power supply of connected units based on several EBC ports, the interface to the ultrasonic sensors, the processing of the motion sensor inputs (bumper), and the control of the drive motors.

The third unit of this segment is the *USB-CAN-Converter Unit*, responsible for the data exchange between the *Embedded PC Unit* and the *Main Control Unit*.

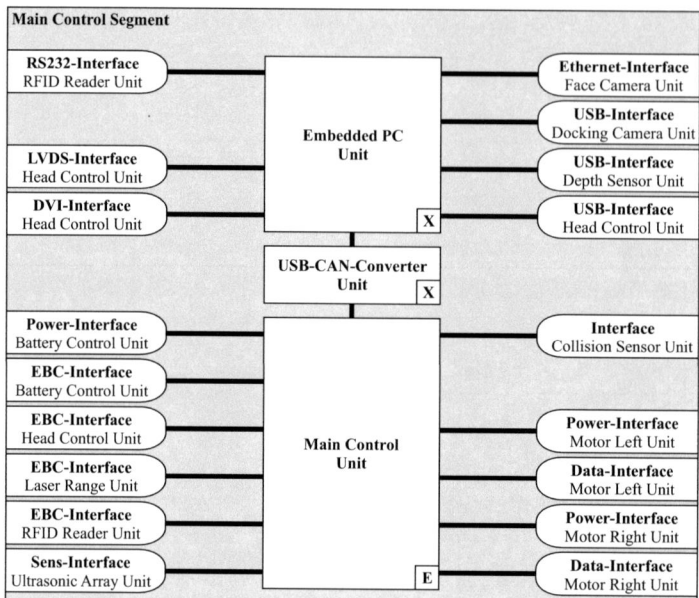

Figure 6.6: *System decomposition of the Main Control Segment (E - Embedded System Unit, H - Hardware Unit, X - External Unit).*

Battery/Charger System Segment

This segment contains all functionalities for the monitoring of the battery and the control of the charging process. The AHP evaluation process (results in Table 6.8) revealed that a lithium battery based on LiFePO$_4$ cells (alternative B4) is most applicable for the home-care robot system. The battery has a nominal voltage of 25.6 V and a nominal capacity of just 40.0 Ah, because of the reduced space inside this robot system. The high number of battery life-cycles, offering significant lower costs over the robot lifetime (see Table 4.4), was the reason why the AHP favored the alternative B4 over B3 used in the shopping robot application.

Table 6.8: *Decision results for the battery systems.*

	A 2.1%	O 11.2%	U 22.6%	R 9.3%	S 14.6%	F 5.0%	C 35.2%	Prio.	Rank
B1	14.8%	14.8%	25.0%	25.0%	25.0%	14.8%	29.4%	**24.7 %**	3
B2	9.0%	9.0%	25.0%	25.0%	25.0%	9.0%	8.4%	**16.2 %**	4
B3	54.6%	54.6%	25.0%	25.0%	25.0%	54.6%	14.4%	**26.7 %**	2
B4	21.6%	21.6%	25.0%	25.0%	25.0%	21.6%	47.8%	**32.4 %**	1

All six charging concepts were evaluated using the AHP (Table 6.9). As a result, the charging concept C4, based on a line voltage autonomous charging system and a line voltage manual charging mode, is most applicable for the integration into this robot system. This decision was mainly influenced by the low costs and the good usability of this alternative.

Table 6.9: *Decision results for the charging systems.*

	A 2.1%	O 11.2%	U 22.6%	R 9.3%	S 14.6%	F 5.0%	C 35.2%	Prio.	Rank
C1	16.7%	7.1%	8.3%	7.1%	13.0%	16.7%	10.0%	**9.9 %**	6
C2	16.7%	7.1%	25.0%	7.1%	13.0%	16.7%	10.0%	**13.7 %**	5
C3	16.7%	35.7%	8.3%	7.1%	5.3%	16.7%	30.0%	**19.1 %**	3
C4	16.7%	35.7%	25.0%	7.1%	5.3%	16.7%	30.0%	**22.8 %**	1
C5	16.7%	7.1%	8.3%	35.7%	31.7%	16.7%	10.0%	**15.3 %**	4
C6	16.7%	7.1%	25.0%	35.7%	31.7%	16.7%	10.0%	**19.1 %**	2

According to both design decisions, the *Battery/Charger System Segment* is decomposed into the units, shown in Figure 6.7. The *Battery Control Unit* is responsible for the monitoring of all battery cell voltages (over-voltage, under-voltage), the balancing of the battery cells, the control of the charging process (setting of voltage and current levels), and the safety shutdown in case of an over-current. Similar to the *Battery Control Unit* of the shopping robot, this unit further executes the power-up procedure of the robot, generates the global system time, and executes a determined power down procedure in case of an empty battery.

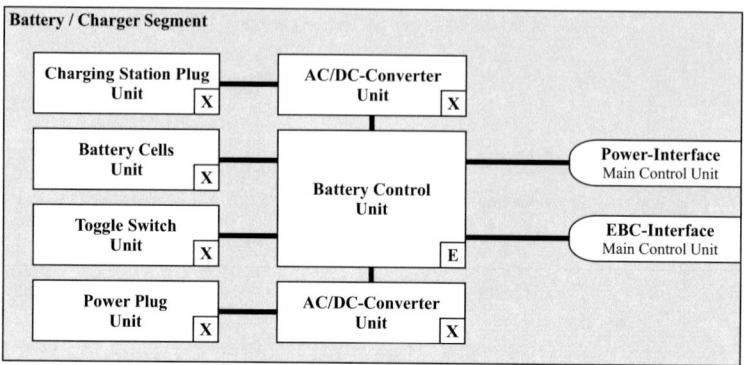

Figure 6.7: *System decomposition of the Battery/Charger Segment (E - Embedded System Unit, X - External Unit).*

Other units of this segment are the *Charging Station Plug Unit* for the contact to the charging station and the *Power Plug Unit* for the connection of a charging cable. The *AC/DC-Converter Units* are applied to convert the incoming line voltage level to an extra-low voltage level usable by the *Battery Control Unit*. The *Toggle Switch Unit* is integrated to turn on and off the robot (e.g., for transportation).

Head/Multimedia Segment

The third segment is the *Head/Multimedia Segment* (Figure 6.8). This segment is responsible for the control of all interactive components (i.e., the touch display, loudspeakers, and microphones) and the control of functionalities supporting the usability of this system (e.g., eye displays, display motor, pet sensor).

The *Head Control Unit* processes the incoming information from the embedded PC and controls all functionalities of the *Head/Multimedia Segment*. Other units are the *Display Unit* and the *Touch Sensor Unit* to present information to the user and to receive user inputs; or the *Display Motor Unit* to tilt the display to an adequate position. The *Eye Display Units* are integrated to indicate the "emotional" state of the robot. The *Pet Sensor Unit* can be used to give the robot a feedback by the user. For sound output and speech input, microphones and loudspeakers are built into this segment.

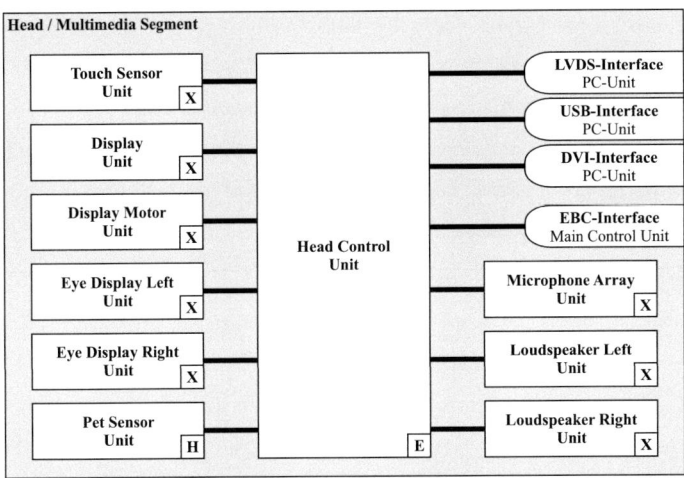

Figure 6.8: *System decomposition of the Head/Multimedia Segment (E - Embedded System Unit, H - Hardware Unit, X - External Unit).*

6.3 Detailed System Design

The detailed system design of this robot focuses on the development of the *Main Control Unit*, the *Battery Control Unit*, and the *Head Control Unit*. The design processes of the *Ultrasonic Sensor Array Unit* and the *Pet Sensor Unit* are not in the scope of this work. All other system units are external units and can be selected from off-the-shelf products.

The hardware and software architectures of the three embedded system units in the focus of this work are similar to the *Robot Head Control Unit* of the shopping robot system (Chapter 5.3). All units consist of the same type of uC for the processing of software algorithms, power supply components for the conversion of the supply voltage, and a CAN interface for the communication with other embedded system units or the embedded PC. The *Main Control Unit* and the *Head Control Unit* further contain an FPGA for the processing of fast signals. The integration of additional unit components depends on the functions of the particular unit (see Section 6.4).

The software architecture includes the Hardware Abstraction Layer (HAL), the *Operation System Layer*, the *Driver Layer*, and the *Application Layer* (Figure 5.12). The communication between system units is also based on the CANopen protocol. Therefore, the *CANopen Layer* with specific *CANopen Object Dictionaries* is part of the software architecture. A detailed description of the system design process of these units is not in the scope of this document.

6.4   System Element Realization

An overview of the *Main Control Unit*, *Battery Control Unit*, and *Head Control Unit*, realized in the course of this work, is presented in the following sections. The evaluation processes of these units (not described in this document) were carried out similar to the *Robot Head Control Unit* of the shopping robot system (Chapter 5.4).

6.4.1   Main Control Unit

The *Main Control Unit* is the co-controller of the robot (Figure 6.9). The tasks of this module are the control of the *Embedded PC Unit*, the interface to the *Battery Control Unit*, the power supply of connected units by EBC ports (e.g., the *Head Control Unit*), the control and power supply of the drive motors, and the interface to the ultrasonic sensors.

For the control of the *Embedded PC Unit*, the *Main Control Unit* includes a power switch to enable or disable the power converter of the embedded PC. The status flags of the embedded PC are monitored. The interface to the *Battery Control Unit* is given by two connectors (battery interfaces): one for the power transmission from the battery, and the other for the exchange of status and communication signals. The integrated EBC ports have the same characteristics as the EBC ports of the shopping robot (voltage and current levels). Every power output is monitored for over-current and can be separately turned on and off. The connected drive motors are controlled by configuration signals (e.g., for directional settings) and PWM signals (for speed settings) generated by the *Main Control Unit*. The power supply of the motors is based on battery voltage. A power switch is integrated to turn-off the motor supply voltage, when the robot is not moving. The current consumption of the motors can be monitored to detect failures of the drive system. The collision sensor of the platform is also connected to the *Main Control Unit*. Hardware and software monitoring provides redundancy and increases the reliability of this safety sensor. The *Main Control Unit* further provides a voltage output and an I2C interface for the connection of ultrasonic sensors.

For the processing of incoming and outgoing information, the *Main Control Unit* is equipped with a uC and an FPGA. The functionalities of both elements can be updated using the CAN bus. The *Main Control Unit* further contains power converters to supply all internal components, and a temperature sensor for the monitoring of the internal system temperature.

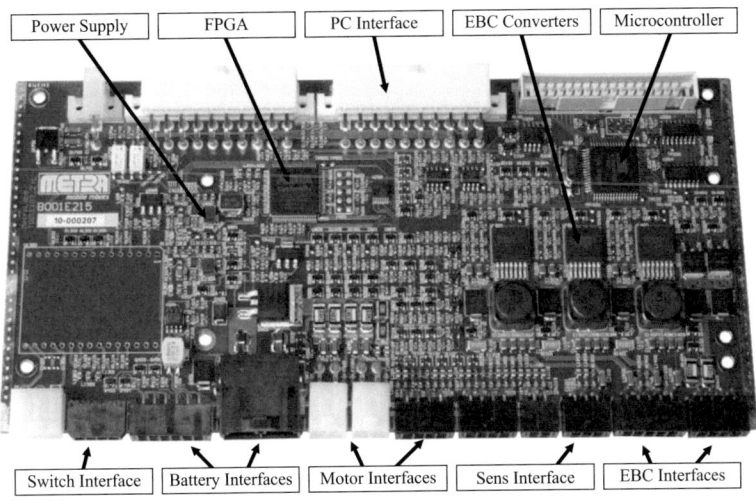

Figure 6.9: *Realization of the Main Control Unit of the home-care robot system.*

6.4.2 Battery Control Unit

This unit handles all functionalities of the battery and the charging system of the robot (Figure 6.10). It is directly connected to all battery cells to monitor the voltage level of each cell. Imbalances of cell voltages are compensated by the integrated balancing switches. All components of the robot system are connected to the battery by the main switch composed of multiple MOSFETs. In case of a cell under- or over-voltage, the integrated uC of the *Battery Control Unit* deactivates this main switch to protect the battery from damage. Another possibility for an emergency shutdown is an over-current. Therefore, the *Battery Control Unit* is unit monitors the charge and discharge currents of the battery to deactivate the main switch as soon as one current exceeds the 20 A threshold.

The integrated uC is further responsible for the control of the battery's charging process. The voltage levels of both charging inputs (for autonomous and manual charging) are continuously monitored for the correct input voltage. The charging process starts as soon as the correct voltage level is provided. Two power converters convert the input voltage to an output voltage, which is adequate for battery charging. The uC can further set the maximum charging current to values up to

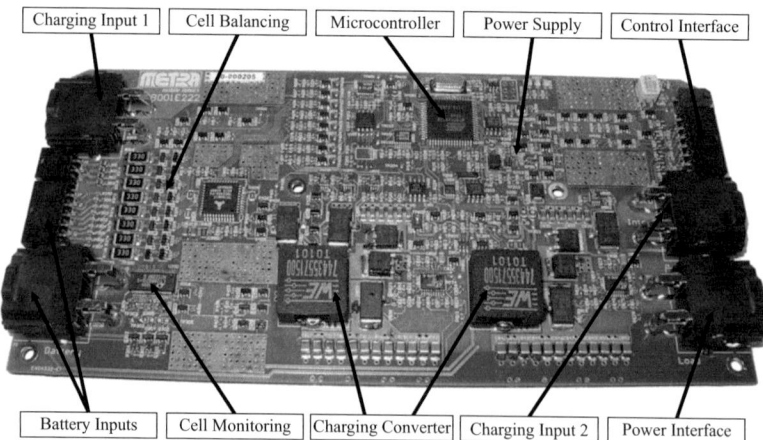

Figure 6.10: *Realization of the Battery Control Unit of the home-care robot system.*

13.5 A. The *Battery Control Unit* can, therefore, deliver a maximum charging power of 400 W.

Moreover, the uC of the *Battery Control Unit* is responsible for the power-up and power-down sequences of the robot as well as the control of the robot's sleep mode. Several control interfaces (e.g., standby voltage, power enable signals, or standby communication) are available to control these modes. A nonvolatile memory is integrated to save occurring error events of the robot.

6.4.3 Head Control Unit

The *Head Control Unit* provides functionalities for the display, the multimedia system, and the robot eyes (Figure 6.11). It provides the power supply for the display and the possibility to dim or to turn-off the display back light for power saving. This unit further contains a motor controller IC to drive the stepper motor to tilt the display. The integrated USB sound card and the related power amplifier generate the output signals for the two loudspeakers of the robot. For speech input, the *Head Control Unit* is equipped with four microphone amplifiers, connected to the sound card.

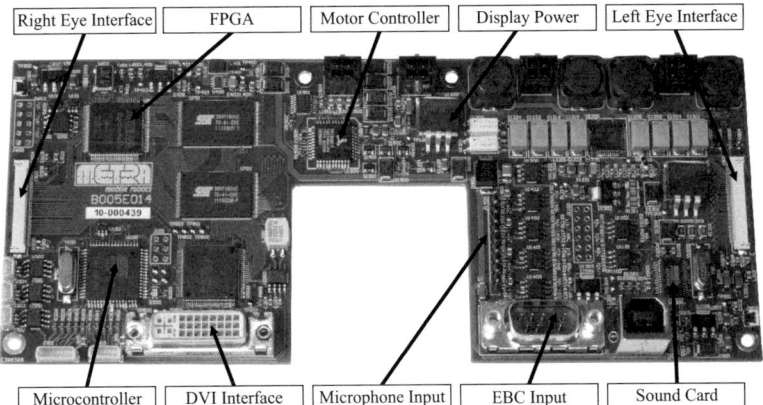

Figure 6.11: *Realization of the Head Control Unit of the home-care robot system.*

The integrated uC is responsible for the control of low-speed signals, i.e., power control signals. To generate the high-speed sequences of the robot eyes, an FPGA is integrated. The eyes are realized by two displays with a resolution of 320x240 pixels and a size of 3.5 inch, and can be used to show different states or emotions of the robot (e.g., sleeping, happy, sad). The pictures of the eyes can be either generated from the integrated memory of the *Head Control Unit* or from the Digital Visual Interface (DVI), generated by the embedded PC.

6.5 System Integration

The designed embedded system units and all selected external units were composed to build all segments of the home-care robot, the subsystems, and finally the overall system. In the following, the composed segments (Figure 6.12) are summarized according to the system decomposition process.

The rubbish bumper is the only unit of the *Motion Sensor Segment* and is placed at the lowest position of the robot base. It determines the maximum step height (1.5 cm) that can be crossed by the robot. The drive system is realized by a differential drive with one castor wheel at the back (version D4 of Figure 4.7). Both, the width and the length of the robot base is about 50 cm. This is significantly smaller than the shopping robot platform with a width of 57 cm and a length of 74 cm. Similar to the shopping robot, the *Drive Motor Segment* of the home-care robot platform consists of two EC motors to move the robot with the required speeds.

All components of the *Battery Charger Segment* are located at low positions of the robot. In particular, the eight lithium-ion cells ($LiFePO_4$) of the battery, which represent about 25 % of the robot's overall weight, are integrated at a very low position. To compensate for the weight distribution of the display and head at the front side of the robot, the battery is moved towards the castor wheel to optimize the stability of the platform. In contrast to the shopping robot, using lithium-polymer cells ($LiCoO_2$) with an integrated Battery Management Unit (BMU), the *Battery Control Unit* of the home-care robot is separated from the battery, which allows

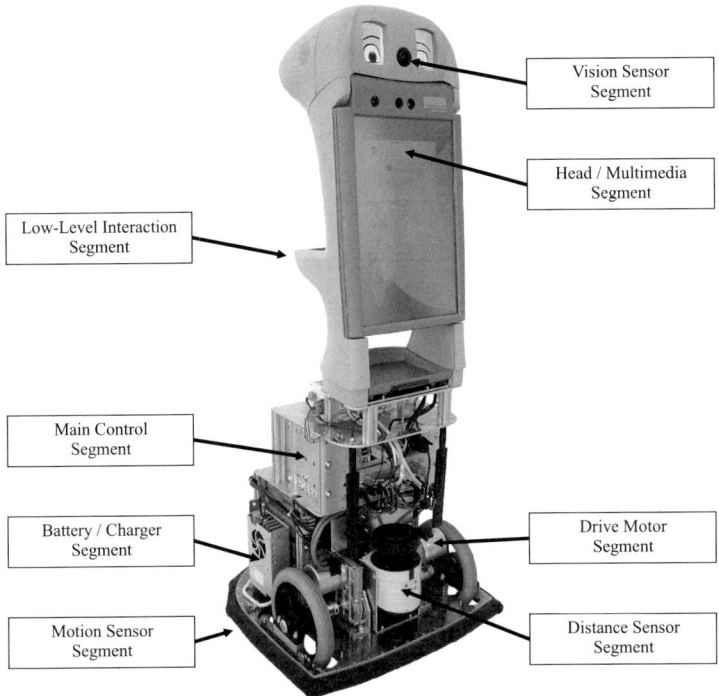

Figure 6.12: *System segments of the finished SCITOS G3 platform.*

flexible component integrations.

The robot can be charged manually using a standard power plug, which can be connected to line power. Different from the shopping robot charging system, the autonomous charging process of the home-care robot is based on line power to safe costs for the charging station. The robot is connected with a power plug socket, known from water boilers, to the autonomous charging station. To compensate for position inaccuracies during the docking process, the charging plug at the station side is mounted on a sliding axle.

The *Distance Sensor Segment* is composed of an S300 laser range finder from the company Sick, which is used for navigation and localization algorithms. Similar to the shopping robot, ultrasonic sensors are used to detect objects with reflective or

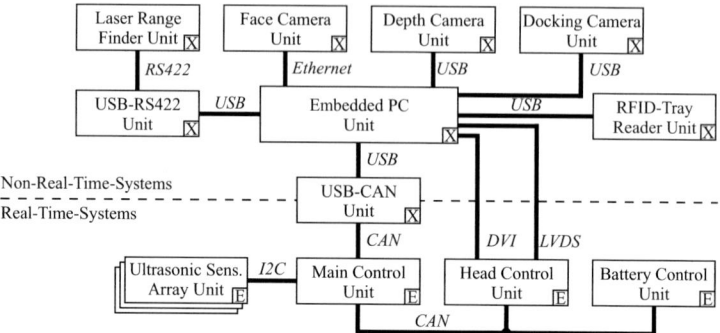

Figure 6.13: *System architecture of the home-care robot system including Embedded System Units (E), External Units (X), and all communication interfaces.*

transparent surfaces, or low heights. Additionally, a depth camera (Kinect sensor from Microsoft) is integrated for the detection of higher obstacles, which might change locations (e.g., chairs or open cabinet doors).

The *Vision Sensor Segment* contains a high-resolution front camera (1600x1200 pixels) in combination with a fish-eye lens, usable for the tracking of persons and for video telephony. A small USB camera supports the identification of the charging station.

For the interaction with users, the *Head/Multimedia Segment* contains a 15.4 inch touch display in portrait format, two loudspeakers inside the robot's head, and an omni-directional microphone on the top of the head. The *Low-Level Interaction Segment* includes the *RFID-Tray Reader Unit*, which can be used to identify labeled objects inside the robot's tray. The tray is positioned to optimize its accessibility for the user.

The *Main Control Segment* consists of an embedded PC with an Intel i7 Core and the *Main Control Unit* described in the previous section. The resulting system architecture with all communication interfaces is shown in Figure 6.13. As determined by the AHP decisions, the system architecture of the home-care robot is less complex compared to the architecture of the shopping robot system.

162

A detailed description of the designed robot system and the arranged evaluation procedures are not in the focus of this work. Further details of the system architecture and the technical realization of the home-care robot system can be found in [Merten and Gross, 2012]. Functional tests of the whole robot system including software algorithms can be found in [Gross et al., 2011b].

6.6 System Delivery and Evaluation

6.6.1 Satisfaction of the AHP Criteria and Evaluation of the Development

Here, the satisfaction of the weighted development criteria for the home-care robot system is summarized. The impact of each criterion (Section 6.1.2) is given in brackets:

Adaptability (2.1 %): The home-care robot is highly specialized for interactive applications in home environments. The embodiment of the robot system (hiding technical details) complicates the integration of novel components. The system architecture, i.e. the embedded system units, is also customized for this particular application. Nevertheless, the usage of the CAN bus and EBC interfaces still allow for a limited integration of further modules (e.g., internal sensors). Therefore, the low-value requirement of system adaptability is satisfied well enough.

Operation Time (11.2 %): The reduced complexity of the system architecture and further optimizations of the components' current consumptions result in an overall power consumption of the robot platform of less than 75 W (in comparison to 100 W of the shopping robot system). Given the battery capacity of 40 Ah and a charging power of 400 W, this allows for operation times with one battery charge of ten to twelve hours (depending on executed software functionalities on the embedded PC). The charging time of an empty battery is about 4.5 h, when all system components are active. The charging time of less than four hours can be achieved by deactivating system components not required during charging (e.g., sensors for navigation and localization) or by slowing-down the computing speed of the embedded PC. Thus, this important criterion is well satisfied and in accordance with the system requirements.

Usability (22.6 %): The usability of the platform is given by the realized drive system that allows for a reliable movement in home environments and the integration of user-friendly charging systems. Components for human-machine interaction were also integrated under consideration of usability aspects: the tiltable touch display can be used in sitting and standing positions, the loud-speakers and microphones are optimized for the interaction in home environments, and the tray is well accessible in standing and sitting positions. The satisfaction of the criterion *Usability* will be evaluated during the test periods.

Robustness (9.3 %): The re-usage of existing software algorithms for navigation and localization, and the taking over of safety concepts from the shopping robot required the integration of industrial components (i.e., laser range finder, embedded PC, face camera, and collision sensor). Therefore, this criterion is highly satisfied. Nevertheless, the high priority of the criterion *Costs* will lead to an exchange of industrial components with low-cost components in further development cycles. This requires the development of novel software and safety concepts.

Safeness (14.6 %): Most of the safety concepts of the home-care robot were taken from the shopping robot system, because of similar requirements of safeness in both application areas. This includes, e.g., the processing of the collision sensor, the analysis of the laser range finder, or the design of the casing. Furthermore, the limited speed of 0.6 m/s increases the safeness of this platform in narrow home environments.

Features (5.0 %): The selection and integration of system components focused on system requirements. Additional features were not included for cost-saving reasons. Further applications of the home-care robot system are restricted to the integrated technologies, e.g., the available computing power or sensor configuration.

Costs (35.2 %): The production costs of this robot are significantly reduced compared to the costs of the shopping robot system, because of a strong focus of the development process on cost-effective solutions. This was mainly achieved by the simplification of the embedded system architecture (control modules and embedded PC), which represents about one third of the costs of the shopping robot. Furthermore, the mechanical design and the casing (about one fifth of the shopping robot costs) were optimized for cost-effective production methods. The integration of a 40 Ah lithium-ion battery (LiFePO$_4$), instead of an 80 Ah lithium-polymer battery (LiCoO$_2$), allowed the reduction of the battery costs by factor three. The costs for the sensor system of the home-care robot (cameras, laser range finder, ultrasonic sensors, and the collision sensor) are similar to the shopping robot. These sensors represent about one third of the production costs of the home-care robot.

The current version of the home-care robot satisfies this criterion already quite well. The estimated costs of the current system are 12.000 Euro. The previously stated requirement of a market prize of less than 10.000 Euro can be realized by an additional development cycle optimizing the costs for the sensor configuration and the control systems.

6.6.2 Field Tests

For the evaluation of the usability of the home-care robot system, two evaluation periods are planned. (The results of both test periods are not available yet.) First tests will be carried out in the test environment provided by a project partner (see Chapter 6.1). In these tests, volunteers suffering from Mild Cognitive Impairment (MCI) live temporarily in the test house where the robot and smart home technologies are installed. Experts for Gerontology of the *CompanionAble* project will supervise and analyze these tests.

In a second evaluation period, the robot system will be applied at the homes of volunteers (Figure 6.14). Here, the robot is planned to operate without supporting smart home systems. This second test period provides a more realistic scenario and

Figure 6.14: *Application scenarios of the home-care robot system. Left picture shows the first prototype, the right picture the second prototype of the home-care robot system.*

should improve the quality of the evaluation.

6.6.3 Market Launch of the Home-Care Robot System

The results of the ongoing field tests will approve the fulfillment of all system requirements and the usability of the developed applications. Depending on these results, a further development cycle might be necessary to integrate further improvements or to correct identified design mistakes. First home-care robots are planned to be commercially available in the middle of 2012.

Chapter 7

Discussion and Conclusion

The design of an interactive service robot is a complex process based on many requirements defined by users, operators, and technical capabilities as well as multiple criteria like costs, usability, reliability, and safeness. The successful realization and marketing of a service robot requires a balanced consideration of all these aspects. The aim of this work was the application and design of systems engineering methods and decision making processes to the developments of two interactive service robots: a shopping robot guiding people in stores to requested product locations and a home-care robot whose task is to assist elderly people to live independently in their home environments. In particular, for the deliberate consideration of all requirements and the individual design criteria of each robot system, this work applied and adapted the V-Model as a system engineering approach for complex development processes. The Analytic Hierarchy Process (AHP) was used to determine optimal design decisions under consideration of weighted criteria for each individual robot system. The combination of both approaches significantly facilitated and improved the development process.

The Application of the V-Model

The developments of both robot systems were based on the V-Model system design approach. The *Project Stages* and *Decision Gates* proposed by the V-Model were highly applicable for the design process of the robots. These stages and gates allowed for a well-structured course of the development; they enforced a deliberate consideration of the system requirements at an early stage of the project, which facilitated the goal-oriented development throughout the project. The simultaneous definition of system elements and the test specifications of these elements ensured the completeness of the development and helped to recognize design mistakes.

The V-Model was originally developed for highly complex military projects. The complexity of the service robot systems is lower compared to these projects. Although all decomposition steps were carried out during the developments of the service robots, this work suggests a slightly simplified course of the original V-Model tasks. The decomposition process should be reduced from four to three abstraction layers omitting the subsystem abstraction layer, because all tasks belonging to this layer can be easily distributed to other decomposition layers. This would simplify the specification afford.

Moreover, the original V-Model applications involve numerous development parties and require a huge amount of documentation. It is important to document the system at the system and unit levels (i.e., functional requirements, non-functional requirements, architectural principles, and evaluations processes). This is advantageous for the communication within the team. The development outputs of the project stages *System Element Realization*, *System Integration*, and *System Delivery* should be documented as suggested by the V-Model to guarantee completeness and reproducibility. In robotics, usually smaller design teams carry out the development of mobile robot systems. Thus, some of the documentation procedures appear redundant. This work suggests to reduce the documentation at the segment level to the specification of evaluation procedures. Other aspects (e.g., functional requirements) are usually already covered by documentations at the other abstraction levels. At unit level, the documentation of units with similar characteristics can be combined and only selected units should be considered for documentation.

The Combination of the V-Model with the AHP Decision Process

At every step of system developments, decisions about technical realizations are required. These decisions are usually made by team members responsible for a particular development state inferred from the total system requirements. Human decision makers are fault-prone, because they often incorporate subjective preferences and goals into the decision, or might neglect criteria important for the overall development goal during decisions at lower abstraction levels. The AHP was applied in this development process to support the decisions of the V-Model for the design of both robots. The AHP required a definition and weighting of criteria, which were important for the final application of each robot. An advantage of the criteria weighting is that the development team has to create a common understanding of the priorities of a development process. Qualitative requirements (e.g., robustness or usability) can be expressed as quantified values of their importance. The AHP is then used to evaluate the alternatives for every decision according to these criteria. This approach facilitated to maintain focused on the entire system goals and ensured the consideration of all requirements in the decisions.

This work has defined seven evaluation criteria and proposes that these criteria can be used in general for the evaluation process of mobile service robots. The criteria have to be weighted specifically for each particular application of a robot system. The outcomes of both development projects validated the usability of these criteria and showed that the different weighting of the criteria leads to individual decision results highly suited for the particular application. Of course, additional criteria or sub-criteria can be defined to further refine and improve the accuracy of AHP design decisions. This work applied the AHP at system, subsystem and segment levels, however, it can also be beneficial for decisions at unit level.

One restriction for the application of the AHP are system requirements that prevent or force the selection of a particular alternative. For example, an expensive laser range finder had to be integrated into the low-cost home-care robot system, because of requirements from navigation and localization algorithms. In such cases, the AHP can be used to estimate the difference of a particular fixed decision compared to an optimal AHP decision. Based on this estimation, the fixed decision might be reconsidered and a development of novel technologies that help to overcome the restriction might be initiated (navigation and localization without laser range finders).

The consideration of system specific characteristics during the weighing of the decision alternatives represents a restriction of the generality of the AHP for different systems. For example, during the weighting of the battery technologies, the available space inside the shopping robot influenced the criteria parameters *Nominal Capacity* and *Costs*. Therefore, the weighting of the battery alternatives is not universally valid and should be re-estimated for every robot system. To overcome this restriction, system independent parameters should be applied, if possible.

One problem of the AHP is the rank reversal problem that might occur when new alternatives are added to the weighting process. For example, the ranking of three alternatives ($A > B > C$) might be changed after the inclusion of a forth alternative D (e.g., $D > B > A > C$). This effect happened during the selection process of the charging technology for the shopping robot system, in which the inclusion of inductive charging systems changed the ranking of the other alternatives. The reasons and possible solutions for the rank reversal problem are described, e.g., in [Barzilai and Golany, 1994].

Comparison of the Applicability of the Developed Robots with Previously Existing Robot Systems

The robot systems described in the state-of-the-art of this document are single installations or announced products. Although the technical characteristics of these robots often met the current state of technology, none of these systems has achieved the awaited breakthrough in the field of service robotics. The major drawbacks of the previously developed shopping robot systems were weak safeness concepts or not sufficient application functionalities; the existing home-care systems were too expensive or had small user benefits. The robot systems developed in this work show similar technical properties; however, components were developed under the careful weighting of design criteria and prioritizing of the requirements, which were important for the market launch as an off-the-shelf product.

This work contributed the embedded systems to the developments of the robots.

The embedded systems represent key technologies for the successful realization of the robots; they include control modules, battery and power management technologies, interaction components, sensor systems, and the drive system. The embedded systems mainly determine the safeness and robustness factors and carry the main proportion of the production costs (about 75% for the shopping robot system).

The safeness of the shopping robot was confirmed by the passing of all verification processes of the German Technical Inspection Agency (TÜV). So far, it is the only autonomous mobile robot for public applications that received the German TÜV certificate. The balanced integration and realization of functionalities allowed further for adequate production costs; more than 50 shopping robots were delivered to various application areas (shopping, guidance, industry, and research).

The home-care robot demonstrates a beneficial combination of reduced production and operation costs with fully autonomous functionalities. The development process and design decisions avoided any technical overhead and considered mandatory requirements searching for cost-effective alternatives. The current version of the home-care robot has an estimated prize of 12.000 Euro. Another development cycle will optimize the production process and allow to further reduce the prize. This remarkable cost-performance ratio, which is an important step to enter the consumer market, has not been achieved by any of the known interactive service robots. The home-care robot relies on a similar safeness concept as the shopping system and will undergo the TÜV certification procedures in the near future. This robot will enter the market of assistive service robots for home environments in the middle of 2012.

Appendix A

Results of the AHP Evaluation Processes

This appendix presents the pairwise comparison results of the AHP evaluation processes for the system architectures, charging systems, and drive systems.

A.1 Evaluation Matrices for the System Architectures

Table A.1: Pairwise comparison results and weights for the system architectures.

Adaptability (A)	A1	A2	A3	A4	Weights	C.R.
A1	1/1	1/1	1/1	1/5	12.5 %	
A2	1/1	1/1	1/1	1/5	12.5 %	0.000
A3	1/1	1/1	1/1	1/5	12.5 %	
A4	5/1	5/1	5/1	1/1	62.5 %	

173

Operating Time (O)	A1	A2	A3	A4	Weights	C.R.
A1	1/1	1/1	5/1	2/1	36.4 %	
A2	1/1	1/1	5/1	2/1	36.4 %	0.010
A3	1/5	1/5	1/1	1/4	6.6 %	
A4	1/2	1/2	4/1	1/1	20.7 %	

Usability (U)	A1	A2	A3	A4	Weights	C.R.
A1	1/1	1/1	1/1	1/1	25.0 %	
A2	1/1	1/1	1/1	1/1	25.0 %	0.000
A3	1/1	1/1	1/1	1/1	25.0 %	
A4	1/1	1/1	1/1	1/1	25.0 %	

Robustness (R)	A1	A2	A3	A4	Weights	C.R.
A1	1/1	1/4	1/5	1/7	5.4 %	
A2	4/1	1/1	1/2	1/5	14.6 %	0.046
A3	5/1	2/1	1/1	1/3	23.7 %	
A4	7/1	5/1	3/1	1/1	56.3 %	

Safeness (S)	A1	A2	A3	A4	Weights	C.R.
A1	1/1	1/5	1/3	1/7	5.7 %	
A2	5/1	1/1	3/1	1/3	26.3 %	0.044
A3	3/1	1/3	1/1	1/5	12.2 %	
A4	7/1	3/1	5/1	1/1	55.8 %	

Features (F)	A1	A2	A3	A4	Weights	C.R.
A1	1/1	1/1	1/5	1/5	8.3 %	
A2	1/1	1/1	1/5	1/5	8.3 %	0.000
A3	5/1	5/1	1/1	1/1	41.7 %	
A4	5/1	5/1	1/1	1/1	41.7 %	

Costs (C)	A1	A2	A3	A4	Weights	C.R.
A1	1/1	1/1	9/1	7/1	43.4%	
A2	1/1	1/1	9/1	7/1	43.4%	0.062
A3	1/9	1/9	1/1	1/4	4.0%	
A4	1/7	1/7	4/1	1/1	9.2%	

A.2 Evaluation Matrices for Charging Systems

Table A.2: Pairwise comparison results and weights for the charging systems.

Adaptability (A)	C1	C2	C3	C4	C5	C6	Weights	C.R.
C1	1/1	1/1	1/1	1/1	1/1	1/1	16.7%	
C2	1/1	1/1	1/1	1/1	1/1	1/1	16.7%	
C3	1/1	1/1	1/1	1/1	1/1	1/1	16.7%	0.000
C4	1/1	1/1	1/1	1/1	1/1	1/1	16.7%	
C5	1/1	1/1	1/1	1/1	1/1	1/1	16.7%	
C6	1/1	1/1	1/1	1/1	1/1	1/1	16.7%	
Operating Time (O)	C1	C2	C3	C4	C5	C6	Weights	C.R.
C1	1/1	1/1	1/5	1/5	1/1	1/1	7.1%	
C2	1/1	1/1	1/5	1/5	1/1	1/1	7.1%	
C3	5/1	5/1	1/1	1/1	5/1	5/1	35.7%	0.000
C4	5/1	5/1	1/1	1/1	5/1	5/1	35.7%	
C5	1/1	1/1	1/5	1/5	1/1	1/1	7.1%	
C6	1/1	1/1	1/5	1/5	1/1	1/1	7.1%	

Usability (U)	C1	C2	C3	C4	C5	C6	Weights	C.R.
C1	1/1	1/3	1/1	1/3	1/1	1/3	8.3%	
C2	3/1	1/1	3/1	1/1	3/1	1/1	25.0%	
C3	1/1	1/3	1/1	1/3	1/1	1/3	8.3%	0.000
C4	3/1	1/1	3/1	1/1	3/1	1/1	25.0%	
C5	1/1	1/3	1/1	1/3	1/1	1/3	8.3%	
C6	3/1	1/1	3/1	1/1	3/1	1/1	25.0%	

Robustness (R)	C1	C2	C3	C4	C5	C6	Weights	C.R.
C1	1/1	1/1	1/1	1/1	1/5	1/5	7.1%	
C2	1/1	1/1	1/1	1/1	1/5	1/5	7.1%	
C3	1/1	1/1	1/1	1/1	1/5	1/5	7.1%	0.000
C4	1/1	1/1	1/1	1/1	1/5	1/5	7.1%	
C5	5/1	5/1	5/1	5/1	1/1	1/1	35.7%	
C6	5/1	5/1	5/1	5/1	1/1	1/1	35.7%	

Safeness (S)	C1	C2	C3	C4	C5	C6	Weights	C.R.
C1	1/1	1/1	3/1	3/1	1/3	1/3	13.0%	
C2	1/1	1/1	3/1	3/1	1/3	1/3	13.0%	
C3	1/3	1/3	1/1	1/1	1/5	1/5	5.3%	0.000
C4	1/3	1/3	1/1	1/1	1/5	1/5	5.3%	
C5	3/1	3/1	5/1	5/1	1/1	1/1	31.7%	
C6	3/1	3/1	5/1	5/1	1/1	1/1	31.7%	

Features (F)	C1	C2	C3	C4	C5	C6	Weights	C.R.
C1	1/1	1/1	1/1	1/1	1/1	1/1	16.7%	
C2	1/1	1/1	1/1	1/1	1/1	1/1	16.7%	
C3	1/1	1/1	1/1	1/1	1/1	1/1	16.7%	0.000
C4	1/1	1/1	1/1	1/1	1/1	1/1	16.7%	
C5	1/1	1/1	1/1	1/1	1/1	1/1	16.7%	
C6	1/1	1/1	1/1	1/1	1/1	1/1	16.7%	

Costs (C)	C1	C2	C3	C4	C5	C6	Weights	C.R.
C1	1/1	1/1	1/3	1/3	1/1	1/1	10.0 %	
C2	1/1	1/1	1/3	1/3	1/1	1/1	10.0 %	
C3	3/1	3/1	1/1	1/1	3/1	3/1	30.0 %	0.000
C4	3/1	3/1	1/1	1/1	3/1	3/1	30.0 %	
C5	1/1	1/1	1/3	1/3	1/1	1/1	10.0 %	
C6	1/1	1/1	1/3	1/3	1/1	1/1	10.0 %	

A.3 Evaluation Matrices for Drive Systems

Table A.3: *Pairwise comparison results and weights for the drive systems.*

Adaptability (A)	D1	D2	D3	D4	D5	D6	Weights	C.R.
D1	1/1	1/1	1/1	1/1	1/1	1/1	16.7 %	
D2	1/1	1/1	1/1	1/1	1/1	1/1	16.7 %	
D3	1/1	1/1	1/1	1/1	1/1	1/1	16.7 %	0.000
D4	1/1	1/1	1/1	1/1	1/1	1/1	16.7 %	
D5	1/1	1/1	1/1	1/1	1/1	1/1	16.7 %	
D6	1/1	1/1	1/1	1/1	1/1	1/1	16.7 %	

Operating Time (O)	D1	D2	D3	D4	D5	D6	Weights	C.R.
D1	1/1	1/5	1/5	1/5	1/5	1/5	3.9 %	
D2	5/1	1/1	1/1	1/1	1/1	1/1	19.2 %	
D3	5/1	1/1	1/1	1/1	1/1	1/1	19.2 %	0.000
D4	5/1	1/1	1/1	1/1	1/1	1/1	19.2 %	
D5	5/1	1/1	1/1	1/1	1/1	1/1	19.2 %	
D6	5/1	1/1	1/1	1/1	1/1	1/1	19.2 %	

Usability (U)	D1	D2	D3	D4	D5	D6	Weights	C.R.
D1	1/1	1/1	1/1	3/1	7/1	7/1	27.0 %	
D2	1/1	1/1	1/1	3/1	7/1	7/1	27.0 %	
D3	1/1	1/1	1/1	3/1	7/1	7/1	27.0 %	0.016
D4	1/3	1/3	1/3	1/1	5/1	5/1	12.1 %	
D5	1/7	1/7	1/7	1/5	1/1	1/1	3.5 %	
D6	1/7	1/7	1/7	1/5	1/1	1/1	3.5 %	

Robustness (R)	D1	D2	D3	D4	D5	D6	Weights	C.R.
D1	1/1	3/1	5/1	3/1	5/1	7/1	42.0 %	
D2	1/3	1/1	3/1	1/1	3/1	5/1	19.2 %	
D3	1/5	1/3	1/1	1/3	1/1	3/1	8.0 %	0.023
D4	1/3	1/1	3/1	1/1	3/1	5/1	19.2 %	
D5	1/5	1/3	1/1	1/3	1/1	3/1	8.0 %	
D6	1/7	1/5	1/3	1/5	1/3	1/1	3.7 %	

Safeness (S)	D1	D2	D3	D4	D5	D6	Weights	C.R.
D1			$R_{ST/PL} = 0.00$				0.0 %	
D2			$R_{ST/PL} = 0.13$				9.5 %	
D3			$R_{ST/PL} = 0.26$				19.0 %	0.000
D4			$R_{ST/PL} = 0.22$				16.1 %	
D5			$R_{ST/PL} = 0.35$				25.5 %	
D6			$R_{ST/PL} = 0.41$				29.9 %	

Features (F)	D1	D2	D3	D4	D5	D6	Weights	C.R.
D1	1/1	1/1	1/1	1/1	1/1	1/1	16.7 %	
D2	1/1	1/1	1/1	1/1	1/1	1/1	16.7 %	
D3	1/1	1/1	1/1	1/1	1/1	1/1	16.7 %	0.000
D4	1/1	1/1	1/1	1/1	1/1	1/1	16.7 %	
D5	1/1	1/1	1/1	1/1	1/1	1/1	16.7 %	
D6	1/1	1/1	1/1	1/1	1/1	1/1	16.7 %	

Costs (C)	D1	D2	D3	D4	D5	D6	Weights	C.R.
D1	1/1	1/1	3/1	1/3	3/1	5/1	19.9 %	
D2	1/1	1/1	3/1	1/3	3/1	5/1	19.9 %	
D3	1/3	1/3	1/1	1/5	1/1	3/1	8.3 %	0.040
D4	3/1	3/1	5/1	1/1	3/1	5/1	38.4 %	
D5	1/3	1/3	1/1	1/3	1/1	3/1	9.2 %	
D6	1/5	1/5	1/3	1/5	1/3	1/1	4.2 %	

Appendix B

Decomposition Elements of the Shopping Robot

Referring to Figure 5.7, this appendix summarizes all remaining decomposition results of system segments to units that were not discussed within Section 5.2.2.

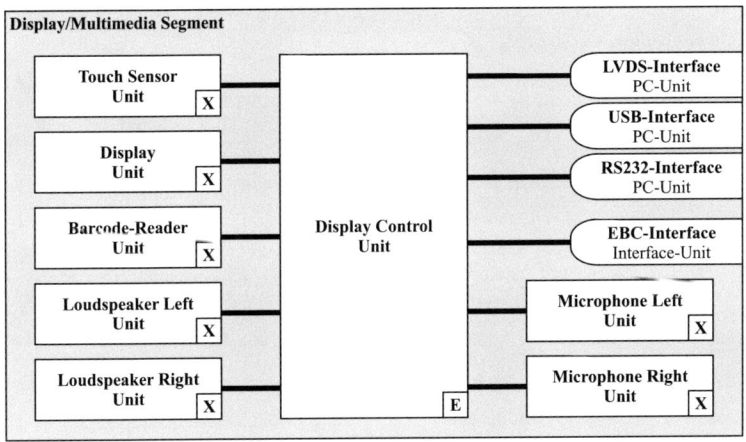

Figure B.1: *System decomposition of the Display/Multimedia Segment.*

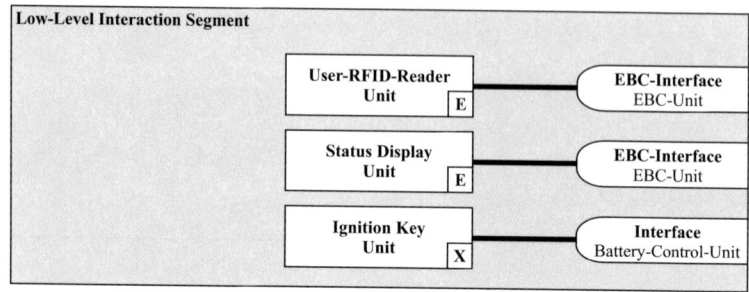

Figure B.2: *System decomposition of the Low-Level Interaction Segment.*

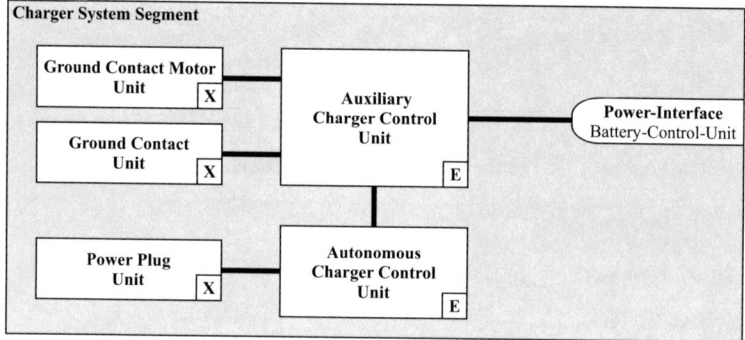

Figure B.3: *System decomposition of the Charger System Segment.*

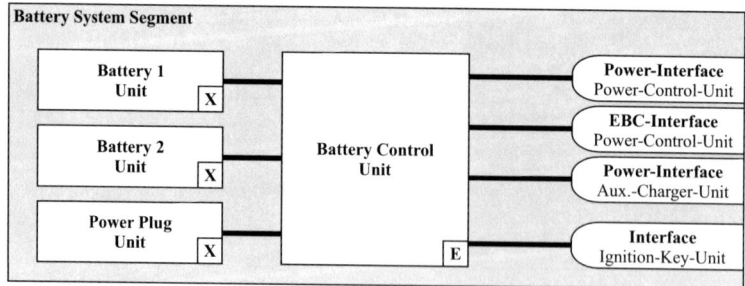

Figure B.4: *System decomposition of the Battery System Segment.*

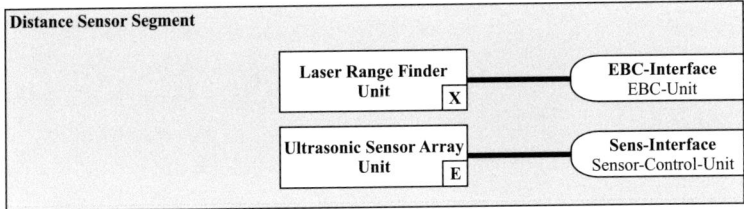

Figure B.5: *System decomposition of the Distance Sensor Segment.*

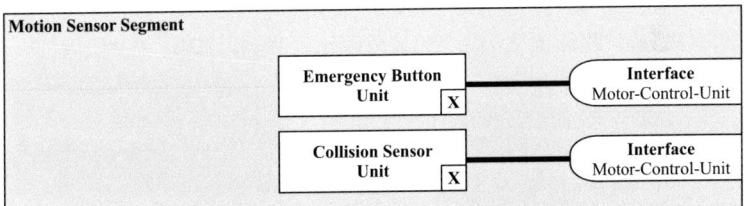

Figure B.6: *System decomposition of the Motion Sensor Segment.*

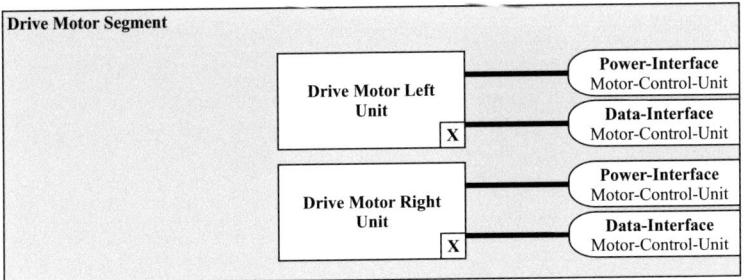

Figure B.7: *System decomposition of the Drive Motor Segment.*

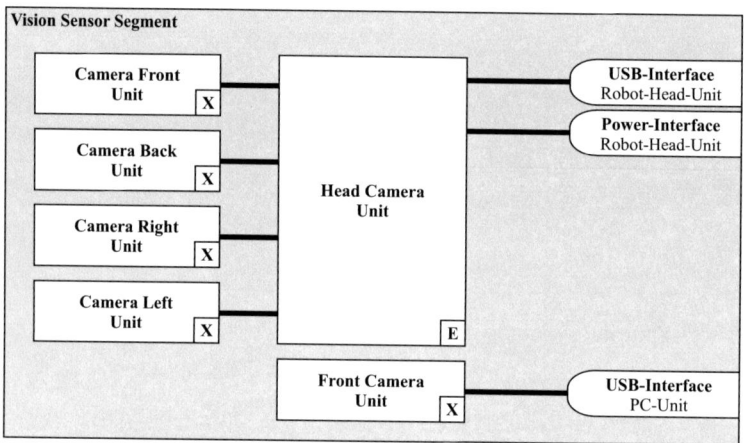

Figure B.8: *System decomposition of the Vision Sensor Segment.*

184

Appendix C

Decomposition Elements of the Home-Care Robot

Referring to Figure 6.5, this appendix summarizes all remaining decomposition results of system segments to units that were not discussed within Section 6.2.2.

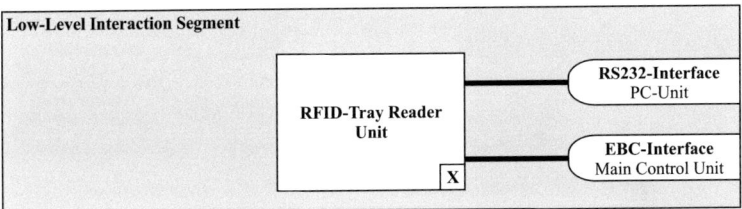

Figure C.1: *System decomposition of the Low-Level Interaction Segment.*

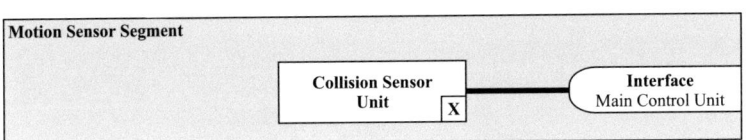

Figure C.2: *System decomposition of the Motion Sensor Segment.*

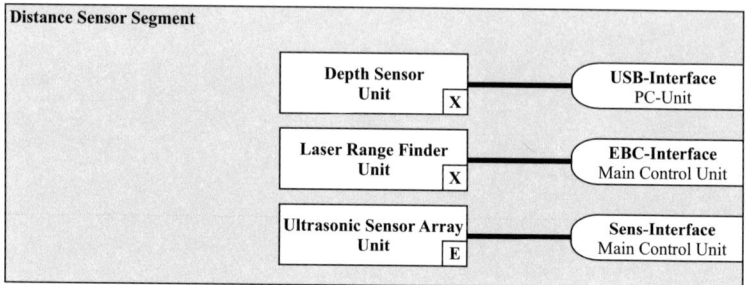

Figure C.3: *System decomposition of the Distance Sensor Segment.*

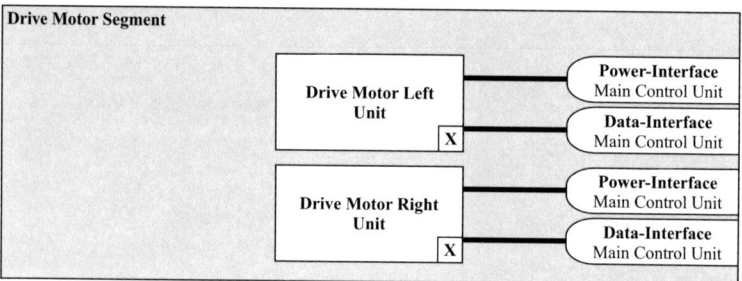

Figure C.4: *System decomposition of the Drive Motor Segment.*

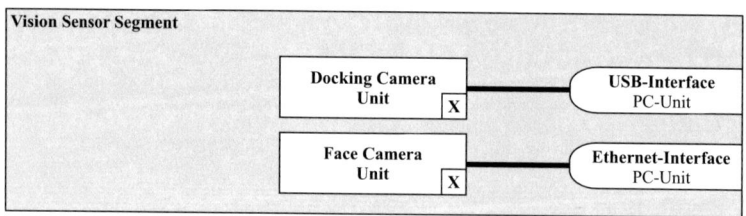

Figure C.5: *System decomposition of the Vision Sensor Segment.*

List of Abbreviations

AAAI   American Association for Artificial Intelligence

ADC   Analog Digital Converter

AGV   Automatic Guided Vehicle

AHP   Analytic Hierarchy Process

BMU   Battery Management Unit

CAN   Controller Area Network

CPU   Central Processing Unit

CG   Care Giver

CR   Care Recipient

CSMA/CA Carrier Sense Multiple Access/Collision Avoidance

DC   Direct Current

DOF   Degrees of Freedom

DVI   Digital Visual Interface

EBC   External Bus Connector

EC   Electronic Commutated

EIA   Electronic Industries Alliance

EMC   Electromagnetic Compatibility

EMI	Electromagnetic Interference
FPGA	Field Programmable Gate Array
GPIO	General Purpose Input/Output
HAL	Hardware Abstraction Layer
HMI	Human Machine Interface
I2C	Inter-Integrated Circuit
IC	Integrated Circuit
ID	Identifier
IFR	International Federation of Robotics
IP-Core	Intellectual Property Core
LED	Light-Emitting Diode
LVDS	Low Voltage Differential Signaling
MCI	Mild Cognitive Impairment
MOSFET	Metal-Oxide-Semiconductor Field-Effect Transistor
Ni-MH	Nickel-Metal Hydride
PC	Personal Computer
PCI	Peripheral Component Interconnect
PCB	Printed Circuit Board
PDO	Process Data Object
PERSES	PERsonenlokalisation und PERsonentracking für mobile SErvicesysteme
PWM	Pulse-Width Modulation
RFID	Radio-Frequency IDentification

RTC	Real Time Clock
SCL	Serial Clock Line
SDA	Serial Data Line
SDLC	System Development Life Cycle
SDO	Service Data Object
SerRoKon	Service-Roboter-Konzeption
SPI	Serial Peripheral Interface
TTL	Transistor-Transistor-Logik
TOF	Time of Flight
uC	Micro Controller
USB	Universal Serial Bus
WLAN	Wireless Local Area Network

List of Figures

Bibliography

[Ackerman and Guizzo, 2011] Ackerman, E. and Guizzo, E. (2011). Mystery Robot Revealed: RoboDynamics Luna Is Fully Programmable Adult-Size Personal Robot. *IEEE Spectrum*. http://spectrum.ieee.org/automaton/robotics/home-robots/robodynamics-luna-fully-programmable-adult-size-personal-robot.

[Ahn et al., 2005] Ahn, S. C., Lee, J.-W., Lim, K.-W., Ko, H., Kwon, Y.-M., and Kim, H.-G. (2005). UPnP Approach for Robot Middleware. In *Proc. IEEE Intern. Conf. on Robotics and Automation (ICRA 2005)*, pages 1971–1975.

[Asfour et al., 2006] Asfour, T., Regenstein, K., Azad, P., Schröder, J., Vahrenkamp, N., and Dillmann, R. (2006). ARMAR-III: An Integrated Humanoid Platform for Sensory-Motor Control. In *Proc. IEEE/RAS International Conference on Humanoid Robots (Humanoids)*, pages 169–175.

[Athawale and Chakraborty, 2011] Athawale, V. M. and Chakraborty, S. (2011). A comparative study on the ranking performance of some multi-criteria decision-making methods for industrial robot selection. *International Journal of Industrial Engineering Computations*, 2:831–850.

[Banga et al., 2007] Banga, V. K., Singh, Y., and Kumar, R. (2007). Simulation of robotic arm using genetic algorithm & AHP. *World Academy of Science, Engineering and Technology*, 25:95–101.

[Barzilai and Golany, 1994] Barzilai, J. and Golany, B. (1994). AHP rank reversal, normalization and aggregation rules. *Information Systems and Operational Research (INFOR Journal)*, 32(2):14–20.

[Berns and Mehdi, 2010] Berns, K. and Mehdi, S. A. (2010). Use of an autonomous mobile robot for elderly care. In *Advanced Technologies for Enhancing Quality of Life (AT-EQUAL)*, pages 121–126.

[Bischoff and Graefe, 2004] Bischoff, R. and Graefe, V. (2004). Hermes a versatile personal robotic assistant. In *Proc. IEEE Special Issue on Human Interactive Robots for Psychological Enrichment*, pages 1759–1779.

[Bluebotics SA, 2011a] Bluebotics SA (2011a). BlueBotics - Entertainment - Gilberto. Website. `http://www.bluebotics.com/entertainment/Gilberto/`.

[Bluebotics SA, 2011b] Bluebotics SA (2011b). Datasheet Gilberto. Website. `http://www.bluebotics.com/entertainment/Gilberto/Gilberto_web.pdf`.

[Boehm, 1988] Boehm, B. W. (1988). A spiral model of software development and enhancement. *Computer*, 21:61–72.

[Britto et al., 2008] Britto, R. S., Santana, A. M., Souza, A. A. S., Medeiros, A. A. D., and Alsina, P. J. (2008). A distributed hardware-software architecture for control an autonomous mobile robot. In *ICINCO-RA (2)*, pages 169–174.

[Burbridge et al., 2011] Burbridge, C., Gatsoulis, Y., and McGinnity, T. M. (2011). Cumulative online unsupervised learning of objects by visual inspection. In *Proc. Conf. on Artificial Intelligence and Cognitive Sciences (AICS 2011)*, pages 86–95.

[Burgard et al., 1999] Burgard, W., Cremers, A. B., Fox, D., Hähnel, D., Lakemeyer, G., Schulz, D., Steiner, W., and Thrun, S. (1999). Experiences with an interactive museum tour-guide robot. *Artificial Intelligence*, 114(1-2):3–55.

[Care-o-Bot, 2011] Care-o-Bot (2011). Care-O-bot 3 Download - Fraunhofer IPA. Website. `http://www.care-o-bot.de/english/Cob3_Download.php`.

[Carnegie et al., 2004] Carnegie, D. A., Prakash, A., Chitty, C., and Guy, B. (2004). A human-like semi autonomous mobile security robot. In *2nd International Conference on Autonomous Robots and Agents (2004)*, pages 64–69.

[Chattopadhyay, 2010] Chattopadhyay, S. (2010). *Embedded System Design*. PHI Learning.

[CommRob, 2009] CommRob (2009). Commrob deliverable D2.2: Final system description.

[CommRob, 2010] CommRob (2010). Commrob deliverable D6.4: Robot trolleys.

[Companionable, 2011] Companionable (2011). Companionable. Website. `http://www.companionable.net`.

[Compaq et al., 2000] Compaq, Hewlett-Packard, Intel, Lucent, Microsoft, NEC, and Philips (2000). Universal serial bus specification revision 2.0.

[Dynamis Batterien, 2009] Dynamis Batterien (2009). Lithium-Poly-Line Battery LP9675135.

[Einhorn et al., 2010] Einhorn, E., Schröeter, C., and Gross, H.-M. (2010). Can't take my eye off you: Attention-driven monocular obstacle detection and 3d mapping. In *Proc. IEEE/RJS Intern. Conf. on Intelligent Robots and Systems (IROS 2010)*, pages 816–821.

[Emmerich, 2011] Emmerich (2011). Specification for sealed rechargeable nickel metal hydride battery. model: EMMERICH NIMH AKKU D 9000 MAH FT-1Z (255047).

[Fraunhofer IPA, 2010] Fraunhofer IPA (2010). Robot systems. Website. `http://www.ipa.fraunhofer.de/index.php?id=17&L=2`.

[Fraunhofer IPA, 2011] Fraunhofer IPA (2011). Museumsroboter Berlin - Fraunhofer IPA. Website. `http://www.ipa.fraunhofer.de/index.php?id=510`.

[Fuchs et al., 2009] Fuchs, M., Borst, C., Giordano, P. R., Baumann, A., Krämer, E., Langwald, J., Gruber, R., Seitz, N., Plank, G., Kunze, K., Burger, R., Schmidt, F., Wimböck, T., and Hirzinger, G. (2009). Rollin' justin - design considerations and realization of a mobile platform for a humanoid upper body. In *Proc. IEEE Intern. Conf. on Robotics and Automation (ICRA 2009)*, pages 4131–4137.

[General Electronics Battery, 2008] General Electronics Battery (2008). Comparison of different battery technologies.

[GPS GmbH, 2011] GPS GmbH (2011). Datenblatt ME-470. Website.

[Graf and Barth, 2002] Graf, B. and Barth, O. (2002). Entertainment robotics: Examples, key technologies and perspectives. In *IROS-Workshop Robots in Exhibitions*.

[Graf et al., 2004] Graf, B., Hans, M., and Schraft, R. D. (2004). Mobile robot assistants. *Robotics & Automation Magazine, IEEE*, 11(2):67–77.

[Graf et al., 2000] Graf, B., Schraft, R. D., and Neugebauer, J. (2000). A mobile robot platform for assistance and entertainment.

[Gross and Böhme, 2000] Gross, H.-M. and Böhme, H.-J. (2000). PERSES - a vision-based interactive mobile shopping assistant. In *Proc. IEEE Int. Conference on Systems, Man and Cybernetics (SMC2000)*, pages 80–85.

[Gross et al., 2009] Gross, H.-M., Böhme, H.-J., Schröeter, C., Müller, S., König, A., Einhorn, E., Martin, C., Merten, M., and Bley, A. (2009). TOOMAS: Interactive shopping guide robots in everyday use - final implementation and experiences from long-term field trials. In *Proc. IEEE/RJS Intern. Conf. on Intelligent Robots and Systems (IROS 2009)*, pages 2005–2012.

[Gross et al., 2008] Gross, H.-M., Böhme, H.-J., Schröeter, C., Müller, S., König, A., Martin, C., Merten, M., and Bley, A. (2008). ShopBot: Progress in developing an interactive mobile shopping assistant for everyday use. In *Proc. 2008 IEEE Int. Conf. on Systems, Man and Cybernetics (SMC 2008)*, pages 3471–3478.

[Gross et al., 2011a] Gross, H.-M., Schröter, C., Müller, S., Volkhardt, M., Einhorn, E., Bley, A., Martin, C., Langner, T., and Merten, M. (2011a). I'll keep an eye on you: Home robot companion for elderly people with cognitive impairment. In *Proc. 2011 IEEE Int. Conf. on Systems, Man and Cybernetics (SMC 2011)*, pages 2481–2488.

[Gross et al., 2011b] Gross, H.-M., Schröter, C., Müller, S., Volkhardt, M., Einhorn, E., Bley, A., Martin, C., Langner, T., and Merten, M. (2011b). Progress in developing a socially assistive mobile home robot companion for the elderly with mild cognitive impairment. In *Proc. IEEE/RJS Intern. Conf. on Intelligent Robots and Systems (IROS 2011)*, pages 2430–2437.

[Heinrich, 2007] Heinrich, G. (2007). *Allgemeine Systemanalyse*. Oldenburg Wissenschaftsverlag GmbH.

[Hosoda et al.,] Hosoda, Y., Egawa, S., Tamamoto, J., Yamamoto, K., Nakamura, R., and Togami, M. In *Proc. IEEE/RJS Intern. Conf. on Intelligent Robots and Systems (IROS 2006)*.

[IFR, 2004] IFR (2004). World Robotics 2004 - Statistics, Market Analysis, Forecasts, Case Studies and Profitability of Robot Investment.

[IFR, 2010] IFR (2010). Service Robots - IFR International Federation of Robotics. Website. `http://www.ifr.org/service-robots/`.

[Infineon, 2011] Infineon (2011). Infineon Technologies AG. Website. `http://www.infineon.com`.

[Jensen et al., 2002] Jensen, B., Philippsen, R., Froidevaux, G., and Siegwart, R. (2002). Mensch-Maschine Interaktion auf der Expo.02.

[Kim et al., 2005] Kim, K. H., Choi, H. D., Yoon, S., Lee, K. W., Ryu, H. S., K.Woo, C., and Kwak, Y. K. (2005). Development of docking system for mobile robots using cheap infrared sensors. In *1st International Conference on Sensing Technology, New Zealand*, pages 287–291.

[Kraft et al., 2010] Kraft, F., Kilgour, K., Saam, R., Stüker, S., Wölfel, M., Asfour, T., and Waibel, A. (2010). Towards social integration of humanoid robots by conversational concept learning. In *Proc. IEEE-RAS Intern. Conf. on Humanoid Robots (Humanoids)*, pages 352–357.

[Kuo et al., 2008] Kuo, T., Broadbent, E., and MacDonald, B. (2008). Designing a robotic assistant for healthcare applications. *The 7th Conference of Health Informatics New Zealand, Rotorua*.

[LitePower Solutions, 2011] LitePower Solutions (2011). Specification of GBS-LFMP60AH.

[Lohmeier et al., 2004] Lohmeier, S., Löffler, K., Gienger, M., Ulbrich, H., and Pfeiffer, F. (2004). Computer System and Control of Biped "Johnnie". In *Proc. IEEE Intern. Conf. on Robotics and Automation (ICRA 2004)*, pages 4222–4227.

[Meeussen et al., 2010] Meeussen, W., Wise, M., Glaser, S., Chitta, S., McGann, C., Mihelich, P., Marder-Eppstein, E., Muja, M. C., Eruhimov, V., Foote, T., Hsu, J., Rusu, R. B., Marthi, B., Bradski, G. R., Konolige, K., Gerkey, B. P., and Berger, E. (2010). Autonomous door opening and plugging in with a personal robot. In *Proc. IEEE Intern. Conf. on Robotics and Automation (ICRA 2010)*, pages 729–736.

[Meixner and Rainer, 2002] Meixner, O. and Rainer, H. (2002). *Computergestütze Entscheidungsfindung. Expert Choise und AHP - innovative Werkzeuge zur Lösung komplexer Probleme.* Redline Wirtschaft.

[Merten and Gross, 2008] Merten, M. and Gross, H.-M. (2008). Highly adaptable hardware architecture for scientific and industrial mobile robots. In *Proc. IEEE Intern. Conf. on Cybernetics and Intelligent Systems and Robotics, Automation and Mechatronics (CIS-RAM 2008)*, pages 1130–1135.

[Merten and Gross, 2012] Merten, M. and Gross, H.-M. (2012). A mobile robot platform for socially assistive home-care applications. In *Robotik 2012 (submitted)*.

[Metro, 2011] Metro (2011). Metro Group Future Store Initiative. Website. `http://www.future-store.org`.

[Michaud et al., 2009] Michaud, F., Ltourneau, D., Beaudry, E., Frchette, M., Kabanza, F., and Lauria, M. (2009). Iterative design of advanced mobile robots. *International Journal of Computing and Information Technology, Special Issue on Advanced Mobile Robotics*, 4:1–16.

[National Semiconductor, 2008] National Semiconductor (2008). LVDS Owners Manual - Including High-Speed CML and Signal Conditioning, Fourth Edition. National Semiconductor.

[Nehmzow, 2009] Nehmzow, U. (2009). *Robot Behaviour: Design, Description, Analysis and Modelling.* Springer-Verlag London.

[Neuner et al., 2003] Neuner, M., Grosser, H., and Brodbeck, Y. (2003). Mediendienst Oktober 2003 - Roboter: Verkaufstalente bei Opel.

[Nguyen et al., 2004] Nguyen, H. G., Morrell, A. J., Mullens, B. K., Burmeister, A. A., Miles, S., Farrington, C. N., Thomas, A. K., and E, D. W. G. (2004). Segway robotic mobility platform. In *in SPIE Mobile Robots XVII*.

[Opel, FHG, 2010] Opel, FHG (2010). Opel ausstellungsroboter. Website. `http://www.ipa.fraunhofer.de/index.php?id=512`.

[Özgürler et al., 2011] Özgürler, S., Güneri, A. F., Gülsün, B., and Yilmaz, O. (2011). Robot selection for a flexible manufacturing system with AHP and TOPSIS methods. In *15th International Research/Expert Conference "Trends in the Development of Machinery and Associated Technology"*, pages 333–336.

[PAL ROBOTICS S.L, 2010a] PAL ROBOTICS S.L (2010a). Datasheet REEM-H2. Website. `http://www.pal-robotics.com/press-kit/brochures/palrobotics-brochure-reem.pdf`.

[PAL ROBOTICS S.L, 2010b] PAL ROBOTICS S.L (2010b). Pal robotics - modular robotics - service robots - humanoid robots - mobile robots. Website. `http://www.pal-robotics.com/`.

[Panasonic, 2011] Panasonic (2011). Value-regulated lead acid batteries: individual data sheet LC-X1242P/LC-X1242AP. Website. `http://www.http://www.panasonic.com/industrial/includos/pdf/Panasonic_VRLA_LC-X1242P_LC_X1242AP.pdf`.

[Park et al., 2005] Park, E., Kobayashi, L., and Lee, S. Y. (2005). Extensible hardware architecture for mobile robots. In *Proc. IEEE Intern. Conf. on Robotics and Automation (ICRA 2005)*, pages 3084–3089.

[Philips Semiconductors, 2000] Philips Semiconductors (2000). The I2C-Bus specification version 2.1. Philips Semiconductors.

[Radack, 2009] Radack, S. (2009). The System Development Life Cycle (SDLC). Website. `http://csrc.nist.gov/publications/nistbul/april2009_system-development-life-cycle.pdf`.

[Reiser et al., 2009a] Reiser, U., Connette, C. P., Fischer, J., Kubacki, J., Bubeck, A., Weisshardt, F., Jacobs, T., Parlitz, C., Hägele, M., and Verl, A. (2009a). Care-o-bot 3 - creating a product vision for service robot applications by integrating design and technology. In *Proc. IEEE/RJS Intern. Conf. on Intelligent Robots and Systems (IROS 2009)*, pages 1992–1998.

[Reiser et al., 2009b] Reiser, U., Parlitz, C., and Klein, P. (2009b). Care-o-bot 3 - vision of a robot butler. In *HCI-Workshop Beyond Gray Droids: Domestic Robot Design for the 21st Century*.

[Robert Bosch GmbH, 1991] Robert Bosch GmbH (1991). Can specification version 2.0. Stuttgart: Robert Bosch GmbH.

[Robosoft SA, 2011] Robosoft SA (2011). Datasheet Kompai. Website. `http://www.robosoft.com/img/tiny/CP/fichKompaiV100425_ebook.pdf`.

[Royce, 1970] Royce, W. W. (1970). Managing the development of large software systems: concepts and techniques. *Proc. IEEE WESTCON, Los Angeles*, pages 1–9. Reprinted in *Proceedings* of the Ninth International Conference on Software Engineering, March 1987, pp. 328–338.

[Ryan and Coup, 2006] Ryan, M. and Coup, R. (2006). An universal, inductively coupled battery charger for robot power supplies. In *Proc. Australasian Conference on Robotics & Automation (ACRA 2006)*.

[Saaty, 1977] Saaty, T. L. (1977). A scaling method for priorities in hierarchical structures. *Journal of Mathematical Psychology*, 15:57–68.

[Saaty, 1980] Saaty, T. L. (1980). *The Analytic Hierarchy Process: planning, priority setting, resource allocation*. McGraw-Hill International Book Co.

[Saaty, 1994] Saaty, T. L. (1994). How to make a decision: The Analytic Hierarchy Process. *Interfaces*, 24:19–43.

[SchultzeWORKS Designstudio, 2011] SchultzeWORKS Designstudio (2011). Gallery Luna personal telepresence robot. Website. `http://www.schultzeworks.com/gallery/`.

[Siegwart et al., 2003] Siegwart, R., Arras, K. O., Bouabdhalla, S., Burnier, D., Froidevaux, G., Jensen, B., Lorotte, A., Mayor, L., Meisser, M., Piguet, R., Ramel, G., Terrien, G., and Tomatis, N. (2003). Robox at expo.02: A large scale installation of personal robots. *Artificial Intelligence*, 42(3-4):203–222.

[Silverman et al., 2003] Silverman, M. C., Jung, B., Nies, D., and Sukhatme, G. S. (2003). Staying alive longer: Autonomous robot recharging put to the test.

[Smith, 1991] Smith, M. F. (1991). *Software Prototyping: Adoption, Practice, and Management*. Mcgraw Hill Book Co. Ltd.

[Soderstrom, 2008] Soderstrom, M. (2008). *Harding Battery Handbook For Quest Rechargeable Cells and Battery Packs*. Harding Energy Inc.

[Staab, 2009] Staab, H. J. (2009). *Ein Radfahrwerk mit passiver Federung für mobile Roboterassistenten*. Jost-Jetter-Verlag Heimsheim.

[Tomatis et al., 2004] Tomatis, N., Bouabdallah, S., Piguet, R., Terrien, G., Siegwart, R., and Sa, B. (2004). Autonomous navigation and security: A 13000h/3000km case study. In *In Proceedings of the 35th International Symposium on Robotics*.

[Tomatis et al., 2003] Tomatis, N., Terrien, G., Piguet, R., Burnier, D., Bouabdallah, S., Arras, K. O., and Siegwart, R. (2003). Designing a secure and robust mobile interacting robot for the long term. In *Proceedings of the 2003 IEEE International Conference on Robotics & Automation*, pages 4246–4251.

[Trabert, 2006] Trabert, J. F. (2006). Interaktive autonome Roboter als Shopping-Assistenz in Baumärkten Innovation für Service mit Marktpotenzial?

[V-Model XT, 2009] V-Model XT (2009). V-model xt complete 1.3. Website. http://v-modell.iabg.de/dmdocuments/V-Modell-XT-Gesamt-Englisch-V1.3.pdf.

[Vargas et al., 2009] Vargas, H., Zanella, V., Martinez, C., Sánchez, J., Martinez, A., Olmedo, E., Toral, D., Pacheco, C., Solis, F., and Ramirez, E. (2009). Nanisha: Service Robot.

[Versteegen, 2002] Versteegen, G. (2002). *Software-Management: Beherrschung des Lifecycles*. Springer-Verlag Berlin.

[Volosyak et al., 2005] Volosyak, I., Ivlev, O., and Graser, A. (2005). Rehabilitation robot FRIEND II - the general concept and current implementation. In *International Conference on Rehabilitation Robotics*, pages 540–544.

[Vorst et al., 2011] Vorst, P., Koch, A., and Zell, A. (2011). Efficient self-adjusting, similarity-based location fingerprinting with passive UHF RFID. In *Proc. IEEE Intern. Conf. on RFID-Technologies and Applications (RFID-TA)*, pages 160–167.

[Wada and Shibata, 2008] Wada, K. and Shibata, T. (2008). Social and physiological influences of living with seal robots in an elderly care house for two months. *Gerontechnology*, 7(2).

[Zell, 2011] Zell, A. (2011). University of Tübingen, Department of Cognitive Systems. Website. `http://http://www.ra.cs.uni-tuebingen.de`.

[Zeltwanger, 2008] Zeltwanger, H. (2008). *CANopen: Das standardisierte, eingebettete Netzwerk*. Vde-Verlag.

[Zou et al., 2006] Zou, D., Wang, T., and Liang, J. (2006). Reconfiguration research on modular mobile robot. In *Proc. IEEE/RJS Intern. Conf. on Intelligent Robots and Systems (IROS 2006)*, pages 1082–1085.

i want morebooks!

Buy your books fast and straightforward online - at one of world's fastest growing online book stores! Environmentally sound due to Print-on-Demand technologies.

Buy your books online at
www.get-morebooks.com

Kaufen Sie Ihre Bücher schnell und unkompliziert online – auf einer der am schnellsten wachsenden Buchhandelsplattformen weltweit! Dank Print-On-Demand umwelt- und ressourcenschonend produziert.

Bücher schneller online kaufen
www.morebooks.de

VDM Verlagsservicegesellschaft mbH
Heinrich-Böcking-Str. 6-8   Telefon: +49 681 3720 174   info@vdm-vsg.de
D - 66121 Saarbrücken   Telefax: +49 681 3720 1749   www.vdm-vsg.de

MIX
Papier aus verantwortungsvollen Quellen
Paper from responsible sources
FSC® C105338

Printed by Books on Demand GmbH, Norderstedt / Germany